Integral Transformations and Anticipative Calculus for Fractional Brownian Motions

MEMOIRS

of the
American Mathematical Society

Number 825

Integral Transformations and Anticipative Calculus for Fractional Brownian Motions

Yaozhong Hu

May 2005 • Volume 175 • Number 825 (first of 4 numbers) • ISSN 0065-9266

American Mathematical Society
Providence, Rhode Island

2000 *Mathematics Subject Classification.*
Primary 60H05, 60H07, 60G30, 60G15, 26A33, 44A05.

Library of Congress Cataloging-in-Publication Data

Hu, Yaozhong, 1961–
    Integral transformations and anticipative calculus for fractional Brownian motions / Yaozhong Hu.
        p. cm. — (Memoirs of the American Mathematical Society, ISSN 0065-9266 ; no. 825)
    "Volume 175, number 825 (first of 4 numbers)."
    Includes bibliographical references.
    ISBN 0-8218-3704-4 (alk. paper)
        1. Stochastic integrals. 2. Gaussian processes. 3. Fractional calculus. 4. Integral transforms. I. Title. II. Series.

QA3.A57   no. 825
[QA274.22]
510 s—dc22
[519.2′2]   2005041980

Memoirs of the American Mathematical Society

This journal is devoted entirely to research in pure and applied mathematics.

Subscription information. The 2005 subscription begins with volume 173 and consists of six mailings, each containing one or more numbers. Subscription prices for 2005 are $606 list, $485 institutional member. A late charge of 10% of the subscription price will be imposed on orders received from nonmembers after January 1 of the subscription year. Subscribers outside the United States and India must pay a postage surcharge of $31; subscribers in India must pay a postage surcharge of $43. Expedited delivery to destinations in North America $35; elsewhere $130. Each number may be ordered separately; *please specify number* when ordering an individual number. For prices and titles of recently released numbers, see the New Publications sections of the *Notices of the American Mathematical Society.*
    Back number information. For back issues see the *AMS Catalog of Publications.*
    Subscriptions and orders should be addressed to the American Mathematical Society, P. O. Box 845904, Boston, MA 02284-5904, USA. *All orders must be accompanied by payment.* Other correspondence should be addressed to 201 Charles Street, Providence, RI 02904-2294, USA.

    Memoirs of the American Mathematical Society is published bimonthly (each volume consisting usually of more than one number) by the American Mathematical Society at 201 Charles Street, Providence, RI 02904-2294, USA. Periodicals postage paid at Providence, RI. Postmaster: Send address changes to Memoirs, American Mathematical Society, 201 Charles Street, Providence, RI 02904-2294, USA.

Contents

Abstract

This paper studies two types of integral transformation associated with fractional Brownian motion. They are applied to construct approximation schemes for fractional Brownian motion by polygonal approximation of standard Brownian motion. This approximation is the best in the sense that it minimizes the mean square error. The rate of convergence for this approximation is obtained. The integral transformations are combined with the idea of *probability structure preserving mapping* introduced in [**48**] and are applied to develop a stochastic calculus for fractional Brownian motions of all Hurst parameter $H \in (0, 1)$. In particular we obtain Radon-Nikodym derivative of nonlinear (random) translation of fractional Brownian motion over finite interval, extending the results of [**48**] to general case. We obtain an integration by parts formula for general stochastic integral and an Itô type formula for some stochastic integral. The conditioning, Clark derivative, continuity of stochastic integral are also studied. As an application we study a linear quadratic control problem, where the system is driven by fractional Brownian motion.

AMS 2000 subject classifications. 60H05, 60H07, 60G30, 60G15, 26A33, 44A05.

Key words and phrases: fractional Brownian motion, integro-differential transformation, probability structure preserving, approximation, stochastic integral, Meyer's inequality, L_p estimate, Itô formula, integration by parts formula, Clark derivative, nonlinear translation, Radon-Nikodym derivative, conditioning, continuous modification, linear quadratic control, Ricatti equation.

This work is supported in part by the National Science Foundation under Grant No. DMS 0204613 and No. EPS-9874732, matching support from the State of Kansas and General Research Fund of the University of Kansas.

Received by the editor November 21, 2002 and in revised form on April 11, 2004.

CHAPTER 1

Introduction

Let $B^H = (B^H_t, t \geq 0)$ be a *fractional Brownian motion* (fBm) with Hurst parameter $H \in (0,1)$. Namely, $(B^H_t, t \geq 0)$ is a centered (mean zero) Gaussian process whose covariance is given by

(1.1) $$\mathbb{E}\left(B^H_t B^H_s\right) = \frac{1}{2}(t^{2H} + s^{2H} - |t-s|^{2H}), \quad s,t \geq 0.$$

This process is *self-similar* in the sense that $(B_{at}, 0 \leq t < \infty)$ has the same probability law as $(a^H B_t, 0 \leq t < \infty)$. It displays a long range dependence and positive correlation properties when $1/2 < H < 1$ and it displays negative correlation property when $0 < H < 1/2$. When $H = 1/2$, it becomes a *standard Brownian motion* and we denote $B^{1/2}_t$ by B_t.

Except when $H = 1/2$, the fractional Brownian motion is not semimartingale and the study of this processes is more difficult and challenging. This makes the investigation of the fractional Brownian motions fascinating. On the other hand and more importantly, since a fractional Brownian motion is not a semimartingale (when $H \neq 1/2$) it may be applied to describe natural or social phenomena that semimartingales fail to fit. This means that the fBm extends the applicability of mathematics (in particular probability). In fact, the fractional Brownian motion has already been successfully applied to hydrology, climatology, signal processing, network traffic analysis, finance as well as various other fields.

Since fractional Brownian motion is not a semimartingale the powerful tool of *stochastic analysis* is not directly applicable. In particular it has been demonstrated that to define a stochastic integral with certain nice properties the integrator process must be a semimartingale. This makes the stochastic analysis for fractional Brownian motion challenging and fascinating. In recent years many authors have devoted to this issue. Let us recall that in the framework of standard Brownian motion there exist two types of stochastic integral:

(1.2) $\quad \int_0^T f(t,\omega)dB_t = \lim\limits_{|\pi|\to 0} \sum\limits_{k=0}^{n-1} f(t_k,\omega)\left(B_{t_{k+1}} - B_{t_k}\right)$ $\qquad$ (Itô)

(1.3)

$\int_0^T f(t,\omega)\delta B_t = \lim\limits_{|\pi|\to 0} \sum\limits_{k=0}^{n-1} \dfrac{f(t_k,\omega) + f(t_{k+1},\omega)}{2}\left(B_{t_{k+1}} - B_{l_k}\right)$ $\quad$ (Stratonovich)

where $\pi : 0 = t_0 < t_1 < \cdots < t_n = T$ is a partition of $[0,T]$ and $|\pi| = \max_{0 \leq k \leq n-1}(t_{k+1} - t_k)$. The first one is called *Itô integral* which produces the famous Itô formula and the second one is called *Stratonovich integral* which yields the usual chain rule. It is natural to try to extend these constructions to fractional Brownian motions. However, as proved in [**35**] (Theorem 3.12) these two

1

constructions yield the same limit if $1/2 < H < 1$. In fact the idea to use (1.2) to define stochastic integral appeared in [25], [74], where $1/2 < H < 1$. Surprisingly, this integral always corresponds to the ordinary chain rule. Along this direction some recent papers have been completed, see [37], [42], [43]. Another important method is Lyons' approach by using the Chen's iterated integrals (see [24], [78], [79] and the references therein). This integration theory can be applied to more general functions and the integrator needs not to be a stochastic process. All these types of stochastic integral lack the mean zero property. Namely, for some f, $\mathbb{E}\left(\int_0^T f(t)\delta B_t^H\right) \neq 0$ (see [35] for examples). This is inconvenient in application: when a stochastic differential system $\dot{x}_t = b(x_t) + \sigma(x_t)\dot{B}_t^H$ is applied to model a certain natural or social system, where x_t is the state of the system, the *drift* term $b(x_t)$ usually represents the mean rate of change and the *diffusion* term $\sigma(x_t)\dot{B}_t^H$ represents the random fluctuation. Since the mean rate of change is completely contained in $b(x_t)$, the random fluctuation $\sigma(x_t)\dot{B}_t^H$ should not contain more average contribution, *i.e.* its mean should be zero. In a rigorous meaning this means that the expectation of the stochastic integral should be zero. In [35] a different type of stochastic integral $\int_0^T f(t)dB_t^H$ is introduced. This is done by replacing the product in (1.2) by the Wick product. This stochastic integral has the property that $\mathbb{E}\left(\int_0^T f(t)dB_t^H\right) = 0$ and the Itô formula takes the form

$$F(B_t^H) = F(B_s^H) + \int_s^t F'(B_r^H)dB_r^H + H\int_s^t r^{2H-1}F''(B_r^H)dr$$

Note that this formula is similar to the standard Itô formula (and formally implies the Itô formula for Brownian motion when H is replaced by $1/2$). For this reason we call $\int_0^T f(t)dB_t^H$ the *Itô stochastic integral* and $\int_0^T f(t)\delta B_t^H$ the *pathwise stochastic integral*. In [35] Malliavin calculus is extensively used. In fact the Malliavin calculus has been applied to fractional Brownian motion in [32], [33]. The stochastic integral $\int_0^t u_s\delta_H B_s^H$ introduced in [33] is different to the one introduced in [35] and to the one in this paper. In fact according to their theorem (Theorem 4.8 of [33]) if u is an adapted process, then $\int_0^t u_s\delta_H B_s^H = \int_0^t u_s dB_s$. The integral $\int_0^T f(t)dB_t^H$ has no such property. Let us also mention the work of stochastic integral for more general Gaussian processes: [2], [3]. A more detailed discussion is presented in Chapter 6.5.

This paper systematically uses the idea of *probability structure preserving* (introduced in [48]) to define stochastic integral. It helps to extend several classical results to fractional Brownian motion case. The idea of probability structure preserving is analogous to the idea of coordinate chart in differential geometry. First a *fractional Wiener functional* (functional of fractional Brownian motion) is transformed into a classical Wiener functional. Some operations are completed for this Wiener functional and then transformed back to the fractional Wiener functional. The role similar to the coordinate chart is played by the operators $\mathbb{B}_H$ and $\mathbb{F}_H$ introduced in Chapter 5.

Now we explain the contents of this paper in detail.

In Chapter 2, we give direct proofs of two representation formulas for fBm (see (2.3), (2.16), and (2.19)). The representation (2.3) has already been known. But no direct proof is available in literature: the proof in [33] uses the analytic continuation and in [90] the representation is obtained by using some intermediate martingales. We give a direct and elementary proof. This proof only requires the use of some integral identities which on its own will be used in other sections of the paper.

In Chapter 3 we introduce and study a class of integro-differential transformation $\mathbb{A}_H$ ((3.1)-(3.2)) associated with one representation of fBm introduced in Chapter 2. In particular, we study the boundedness of $\mathbb{A}_H$. This result will be used to estimate the rate of convergence of the approximation of fBm by polygonal approximation of sBm.

In Chapter 4, we construct approximation of fBm using the polygonal approximation of the sBm. Let $\pi : 0 = t_0 < t_1 < \cdots < t_n = T$ be a partition of the interval $[0, T]$ and let B_t^π be the polygonal approximation of B_t. Combining B_t^π with the representation of fBm we obtain an approximation of $B_t^{H,\pi}$. Using the boundedness results of $\mathbb{A}_H$ obtained in Chapter 3, we prove that $B_t^{H,\pi}$ converges to B_t^H. We also obtain the rate of convergence. More precisely, we obtain that

$$(1.4) \qquad \mathbb{E} \sup_{0 \le t \le T} |B_t^{H,\pi} - B_t^H| \le C_{H,\gamma,T} \left(\frac{1}{n}\right)^\gamma,$$

where

$$(1.5) \qquad \gamma < \begin{cases} H & if \ 0 < H < 1/2 \\ 1 - H & if \ 1/2 < H < 1. \end{cases}$$

Moreover, if $3/4 < H < 1$, then we may design a special type of partition π with the property that it is is finer near 0 than near T. For this partition the rate of convergence γ can be any number satisfying

$$(1.6) \qquad 0 < \gamma < (1 - H)(4H - 2).$$

Since $(1 - H)(4H - 2) > 1 - H$ when $H > 3/4$, this means that this particular choice of partition π can improve the rate of convergence if $H > 3/4$.

It is interesting to note that this approximation of fBm is the *best* in the following sense. Let the standard Brownian motion be given at some sampling points $0 = t_0 < t_1 < \cdots < t_n = T$, namely the random variables $B_{t_1}, \cdots, B_{t_n}$ are known. If we use a function of $B_{t_1}, \cdots, B_{t_n}$, i.e. $g(t, B_{t_1}, \cdots, B_{t_n})$ to approximate B_t^H, then the $B_t^{H,\pi}$ constructed above is the best in the sense that it minimizes

$$(1.7) \qquad \mathbb{E} |g(t, B_{t_1}, \cdots, B_{t_n}) - B_t^H|^2$$

among all possible (measurable) function of g and for all $0 \le t \le T$. Note that this approximation appeared also in [**33**], where the convergence in L^2 norm is studied. Namely it is proved that $\lim_{|\pi| \to 0} \int_0^T \mathbb{E} |B_t^{H,\pi} - B_t^H|^2 dt = 0$. But there is no rate estimate there. Moreover, the convergence studied in this paper is in terms of *sup* norm.

In Chapter 5, we introduce and study another class of integro-differential transformation $\mathbb{F}_{H,T}$ associated with the representation of the fBm introduced in Chapter 2. This transformation plays fundamental role in our definition of stochastic integral, Itô formula, Girsanov type formula, conditioning and so forth. The main results of this section are the explicit forms of the transpose $\mathbb{F}_{H,T}^*$, the inverse $\mathbb{B}_{H,T}$, and the composition of these transformations, i.e. $\mathbb{F}_{H,T}\mathbb{F}_{H,T}^*$, $\mathbb{B}_{H,T}^*\mathbb{B}_{H,T}$.

The Itô stochastic integral is defined and studied in Chapters 6 and 7. First we introduce a probability structure preserving mapping induced by the transformation

$\mathbb{T}_{H,T}$ defined in Chapter 5. Then we define the stochastic integral for fBm by pulling back to the sBm case. It is similar in spirit to the construction of calculus on manifold by the coordinate chart. This way of defining stochastic integral is briefly mentioned in [**48**]. Chapters 6 and 7 give a complete study of this approach. The definition of stochastic integral in this paper is very general. A broad class of stochastic processes is integrable. The condition for the existence of stochastic integral, Meyer's inequality, and L^p estimate of the stochastic integral are obtained by using the idea of probability structure preserving mapping. But as in the usual case it is always difficult to evaluate some interesting stochastic integrals. For this reason we introduce the *algebraically integrable* concept following the idea of creation and annihilation operators in quantum field theory or quantum probability (see [**7**], [**83**] and references therein). This makes a full use of the technique of chaos expansion. This idea can be used to compute some specific stochastic integrals. We can also prove an Itô formula using this idea.

The Girsanov type formula for nonlinear translation of fBm is given in Chapter 8. In [**48**] a Girsanov formula was obtained for fractional Brownian motion over a half line $\mathbb{R}_+$. The Radon-Nikodym derivative depends on the fractional Brownian motion B_t^H for all $t \geq 0$. If we want to consider nonlinear translation of fBm over a finite interval $[0, T]$, then we wish that the Radon-Nikodym derivative depends only on the fBm over $[0, T]$. One way is to take the conditional expectation as suggested in [**48**] (see the discussion on conditional expectation in Chapter 9). However, this is complicated. Several researchers raised this question to the author and this section will present an answer.

The Girsanov type formula has been studied for general Gaussian measure by many authors, see for example, [**71**], [**72**], [**94**], [**109**] and the references therein. The main difficulty is to compute a certain Carleman determinant for a certain transformation in Hilbert space. The explicit form of Carleman determinant is known in the sBm case (see [**17**] and the references therein). In [**33**] there is also a formula (c.f. Theorem 4.9 of the paper) for the Radon-Nikodym density. The difference here is that our results are more explicit. Moreover we do not require the translation to be adapted.

It is clear that the conditioning is an important topic of probability theory. The conditional expectation of linear functional of fractional Brownian motion was explicitly given for instance in [**44**], [**47**]. However, we are unaware of conditional expectation for general nonlinear functional of fBm. In Chapter 9 first we present some alternative formulas to compute the conditional expectation of linear functional using the explicit forms of the transformations obtained in Chapter 5. Since any (square integrable) nonlinear functional can be represented as a chaos expansion, we give formulas for computing conditional expectation of a multiple stochastic integral (of Itô type). We also introduce multiple stochastic integrals of Stratonovich type and compute their conditional expectation. The Hu-Meyer formula ([**29**], [**52**]-[**54**], [**65**], [**66**], [**103**]) has been also extended to fractional Brownian motion.

As is well-known, an Itô formula has been obtained in [**35**] for $H > 1/2$ and an Itô formula obtained in [**3**] can be applied to fBm of Hurst parameter $H > 1/4$. It appears that the Itô formula obtained in [**3**] and the Itô formula obtained in [**35**] are different. But it will be shown in Chapter 11 that when applied to fBm with Hurst parameter $H > 1/2$, they are in fact the same formula.

Chapter 12 concerns with the Clark type formula. It is known that this type of formula is important for example in obtaining the replicating portfolio policy in finance. In [58], a Clark type formula is obtained and applied to option pricing in fractal market. In [60] it has been applied to an optimal consumption and portfolio problem in the fractal market. However, the question of uniqueness has not been concerned. Chapter 12 deals with the uniqueness problem by using the transformation from Chapter 5.

The continuity of stochastic integral as a function of the upper limit is important and studied in sBm case. Chapter 13 gives conditions on $f(s)$ such that $\int_0^t f(s) dB_s^H$ is (almost surely) continuous with respect to t. The main tool is the L^p estimate of stochastic integral obtained in Chapter 7.

As an immediate application of the stochastic calculus developed in this paper we shall solve a stochastic optimal control problem where the utility functional is quadratic and the controlled system is a linear stochastic differential equation driven by a fractional Brownian motion of any Hurst parameter. The optimal control is explicitly obtained by solving a Ricatti type equation.

Some results of this paper may be extended to general Gaussian process by using the *reproducing kernel Hilbert space*. However since we are concerned with the explicit form this aspect of generality is not pursued in this article.

CHAPTER 2

Representations

It is known that the fractional Brownian motion is the *fractional derivative* of standard Brownian motion. More precisely, the fBm $B^H = (B_t^H, t \geq 0)$ with *Hurst parameter* $H \in (0,1)$ admits the following representation:

$$(2.1) \qquad B_t^H = \kappa_H \left\{ \int_0^t (t-u)^{H-\frac{1}{2}} dB_u + \int_{-\infty}^0 \left[(t-u)^{H-\frac{1}{2}} - (-u)^{H-\frac{1}{2}} \right] dB_u \right\},$$

where $B = (B_s, s \in \mathbb{R})$ is a Brownian motion on some probability space $(\Omega, \mathcal{F}, P)$, and

$$(2.2) \qquad \kappa_H = \sqrt{\frac{2H\Gamma(\frac{3}{2} - H)}{\Gamma(H + \frac{1}{2})\Gamma(2 - 2H)}}.$$

(Note that this representation is true even when $H = 1/2$.) The probability space will be fixed and the expectation on $(\Omega, \mathcal{F}, P)$ is denoted by $\mathbb{E}$. Note that our definition is different by a constant to some definitions in literature. Due to their important applications, the fBms have been studied by many authors in recent years. Several kinds of stochastic calculus have been developed (see e.g. [2], [33], [35], [58] and the references therein). In particular, in [58] a white noise calculus was developed and applied to investigate a challenging arbitrage problem in mathematical finance.

Let $\Gamma(\alpha) = \int_0^\infty s^{\alpha-1} e^{-s} ds$, $\lambda > 0$, be the gamma function and $B(\mu, \nu) = \frac{\Gamma(\mu)\Gamma(\nu)}{\Gamma(\mu + \nu)}$, $\mu, \nu > 0$, be the beta function. The following alternative representations for fBm are known (see e.g. [33], [90]). In [33], analytic continuation of hypergeometric function is used to discuss the case $H < 1/2$. In [90], these formulas are obtained by using intermediate martingales. There exists no elementary proof yet. It is useful and interesting to give a direct and elementary proof of the representation. In the proof of the following representation we use only some well-known integration identities.

THEOREM 2.1. *Let $H \in (0,1)$. Define*

$$(2.3) \qquad B_t^H = \int_0^t Z_H(t,s) dB_s, \ 0 \leq t < \infty,$$

where $(B_t, t \geq 0)$ is a standard Brownian motion and
(2.4)
$$Z_H(t,s) = \kappa_H \left[\left(\frac{t}{s}\right)^{H-\frac{1}{2}} (t-s)^{H-\frac{1}{2}} - (H - \frac{1}{2}) s^{\frac{1}{2}-H} \int_s^t u^{H-\frac{3}{2}} (u-s)^{H-\frac{1}{2}} du \right].$$

Then $(B_t^H, 0 \leq t < \infty)$ is a fractional Brownian motion of Hurst parameter H. Moreover, if $H > 1/2$, then $Z_H(t, s)$ can be written as

$$(2.5) \qquad Z_H(t, s) = (H - \frac{1}{2})\kappa_H s^{\frac{1}{2} - H} \int_s^t u^{H - \frac{1}{2}}(u - s)^{H - \frac{3}{2}} du.$$

Proof It suffices to show that $\mathbb{E}(B_t^H) = 0$ and

$$\mathbb{E}(B_t^H B_s^H) = \frac{1}{2}(t^{2H} + s^{2H} - |t - s|^{2H}), \qquad 0 \leq s, t < \infty.$$

First let us consider the case when $H > 1/2$. Let X_t denote the right hand side of (2.3), where $Z_H(t, s)$ is given by (2.5). We shall compute $\mathbb{E}(X_t X_r)$ for $0 \leq r \leq s < \infty$. Set $\alpha = H - \frac{1}{2}$ and use the notation $u \wedge v := \min(u, v)$. Then we have

$$
\begin{aligned}
\mathbb{E}(X_t X_r) &= \int_0^r Z_H(t, s) Z_H(r, s) ds \\
&= \left(H - \frac{1}{2}\right)^2 \kappa_H^2 \int_0^r s^{-2\alpha} \int_s^t \int_s^r u^\alpha v^\alpha (u - s)^{\alpha - 1}(v - s)^{\alpha - 1} du\, dv\, ds \\
(2.6) \qquad &= \left(H - \frac{1}{2}\right)^2 \kappa_H^2 \int_0^t \int_0^r u^\alpha v^\alpha \rho(u, v) du\, dv,
\end{aligned}
$$

where

$$\rho(u, v) := \int_0^{u \wedge v} s^{-2\alpha}(u - s)^{\alpha - 1}(v - s)^{\alpha - 1} ds,$$

which is symmetric with respect to u and v. From Lemma 15.3 it follows that

$$\rho(u, v) = B\left(2 - 2H, H - \frac{1}{2}\right) u^{\frac{1}{2} - H} v^{\frac{1}{2} - H}(v - u)^{2H - 2}.$$

Substituting it into (2.6) yields

$$\mathbb{E}(X_t X_r) = \left(H - \frac{1}{2}\right)^2 \kappa_H^2 B\left(2 - 2H, H - \frac{1}{2}\right) \int_s^t \int_s^r |v - u|^{2H - 2} du\, dv.$$

It is straightforward to verify that

$$\kappa_H^2 \left(H - \frac{1}{2}\right)^2 B\left(2 - 2H, H - \frac{1}{2}\right) = H(2H - 1).$$

Thus

$$
\begin{aligned}
\mathbb{E}(X_t X_r) &= H(2H - 1) \int_s^t \int_s^r |v - u|^{2H - 2} du\, dv \\
&= \frac{1}{2}\left[t^{2H} + r^{2H} - |t - s|^{2H}\right].
\end{aligned}
$$

It is easy to see that $(X_t, t \geq 0)$ is a Gaussian process with mean 0. Thus by definition $(X_t, t \geq 0)$ is an fBm. Now the integration by parts formula yields the expression (2.4) for $Z_H(t, s)$.

The situation that $H < 1/2$ is more complicated. We continue to use the notation $\alpha = H - \frac{1}{2}$. Denote

$$(2.7) \qquad G(s, r) = \int_0^{s \wedge r} u^{-2\alpha}(s - u)^\alpha (r - u)^\alpha du.$$

Let $Y_t = \int_0^t Z_H(t,s)dB_s$, where $Z_H(t,s)$ is given by (2.4). Let $0 < t_1 \le t_2 < \infty$.

$$
\begin{aligned}
\frac{1}{\kappa_H^2}\mathbb{E}\left(Y_{t_1}Y_{t_2}\right) &= \int_0^{t_1} Z_H(t_1,s)Z_H(t_2,s)ds \\
&= t_1^\alpha t_2^\alpha \int_0^{t_1} s^{-2\alpha}(t_1-s)^\alpha(t_2-s)^\alpha ds \\
&\quad -\alpha t_2^\alpha \int_0^{t_1} s^{-2\alpha}(t_2-s)^\alpha \int_s^{t_1} u^{\alpha-1}(u-s)^\alpha du\, ds \\
&\quad -\alpha t_1^\alpha \int_0^{t_1} s^{-2\alpha}(t_1-s)^\alpha \int_s^{t_2} u^{\alpha-1}(u-s)^\alpha du\, ds \\
&\quad +\alpha^2 \int_0^{t_1} s^{-2\alpha} \int_s^{t_1} \int_s^{t_2} u^{\alpha-1}(u-s)^\alpha v^{\alpha-1}(v-s)^\alpha du\, dv\, ds \\
&= t_1^\alpha t_2^\alpha G(t_1,t_2) - \alpha t_2^\alpha \int_0^{t_1} u^{\alpha-1} \int_0^u s^{-2\alpha}(u-s)^\alpha(t_2-s)^\alpha ds \\
&\quad -\alpha t_1^\alpha \int_0^{t_2} u^{\alpha-1} \int_0^{t_1\wedge u} s^{-2\alpha}(t_1-s)^\alpha(u-s)^\alpha ds\, du \\
&\quad +\alpha^2 \int_0^{t_1} \int_0^{t_2} u^{\alpha-1}v^{\alpha-1} \int_0^{u\wedge v} s^{-2\alpha}(u-s)^\alpha(v-s)^\alpha ds\, du\, dv \\
&= t_1^\alpha t_2^\alpha G(t_1,t_2) - \alpha t_2^\alpha \int_0^{t_1} u^{\alpha-1} G(u,t_2)du \\
&\quad -\left[\alpha t_1^\alpha \int_0^{t_2} v^{\alpha-1} G(t_1,v)dv - \alpha^2 \int_0^{t_1}\int_0^{t_2} u^{\alpha-1}v^{\alpha-1}G(u,v)du\, dv\right] \\
&= t_2^\alpha\left[t_1^\alpha G(t_1,t_2) - \alpha\int_0^{t_1} u^{\alpha-1}G(u,t_2)dv\right] \\
&\quad -\alpha\int_0^{t_2} v^{\alpha-1}\left[t_1^\alpha G(t_1,v) - \alpha\int_0^{t_1} u^{\alpha-1}G(u,v)du\right]dv \\
&= t_2^\alpha\int_0^{t_1} u^\alpha \frac{\partial}{\partial u}G(u,t_2)du - \alpha\int_0^{t_2} v^{\alpha-1}\left\{\int_0^{t_1} u^\alpha \frac{\partial}{\partial u}G(u,v)du\right\}dv \\
&= t_2^\alpha\int_0^{t_1} u^\alpha \frac{\partial}{\partial u}G(u,t_2)du - \alpha\int_0^{t_1} u^\alpha\left\{\int_0^{t_2} v^{\alpha-1}\frac{\partial}{\partial u}G(u,v)dv\right\}du \\
&= \int_0^{t_1} u^\alpha\left[t_2^\alpha \frac{\partial}{\partial u}G(u,t_2) - \alpha\int_0^{t_2} v^{\alpha-1}\frac{\partial}{\partial u}G(u,v)dv\right]du \\
&= \int_0^{t_1} u^\alpha \frac{\partial}{\partial u}\left\{t_2^\alpha G(u,t_2) - \alpha\int_0^{t_2} v^{\alpha-1}G(u,v)dv\right\}du \\
&= \int_0^{t_1} u^\alpha \frac{\partial}{\partial u}\left\{\int_0^{t_2} v^\alpha \frac{\partial}{\partial v}G(u,v)dv\right\}du\,. \quad (2.8)
\end{aligned}
$$

Let us compute

$$
(2.9) \qquad\qquad I_1 := \int_0^{t_2} v^\alpha \frac{\partial}{\partial v}G(u,v)dv\,.
$$

First we need to compute $\frac{\partial}{\partial v} G(u, v)$. We divide into two cases: $u < v$ and $u > v$. If $u < v$, then

$$
\begin{aligned}
\frac{\partial}{\partial v} G(u, v) &= \frac{\partial}{\partial v} \int_0^u \xi^{-2\alpha} (u - \xi)^\alpha (v - \xi)^\alpha d\xi \\
&= \alpha \int_0^u \xi^{-2\alpha} (u - \xi)^\alpha (v - \xi)^{\alpha - 1} d\xi \\
&= \alpha \int_0^1 \eta^{-2\alpha} (\frac{v}{u} - \eta)^{\alpha - 1} (1 - \eta)^\alpha d\eta .
\end{aligned}
$$

The last integral can be computed by using Lemma 15.2 with $\mu = 1 - 2\alpha$, $\nu = \alpha + 1$. Thus

(2.10)

$$
\begin{aligned}
\frac{\partial}{\partial v} G(u, v) &= \alpha(1 - \alpha) B(1 - 2\alpha, \alpha + 1) \left(\frac{v}{u}\right)^{-\alpha} \int_0^1 \eta^{-\alpha} \left(\frac{v}{u} - \eta\right)^{2\alpha - 1} d\eta \\
&= \alpha(1 - \alpha) B(1 - 2\alpha, \alpha + 1) v^{-\alpha} \int_0^u \xi^{-\alpha} (v - \xi)^{2\alpha - 1} d\xi .
\end{aligned}
$$

If $u > v$, then

$$
\begin{aligned}
G(u, v) &= \int_0^v \xi^{-2\alpha} (v - \xi)^\alpha (u - \xi)^\alpha d\xi \\
&= v \int_0^1 \eta^{-2\alpha} (1 - \eta)^\alpha \left(\frac{u}{v} - \eta\right)^\alpha d\eta .
\end{aligned}
$$

Thus

$$
\begin{aligned}
\frac{\partial}{\partial v} G(u, v) &= \int_0^1 \eta^{-2\alpha} (1 - \eta)^\alpha \left(\frac{u}{v} - \eta\right)^\alpha d\eta \\
&\quad - \frac{\alpha u}{v} \int_0^1 \eta^{-2\alpha} (1 - \eta)^\alpha \left(\frac{u}{v} - \eta\right)^{\alpha - 1} d\eta \\
(2.11) \qquad &= v^{-1} G(u, v) - \frac{\alpha u}{v} \int_0^v \xi^{-2\alpha} (v - \xi)^\alpha (u - \xi)^{\alpha - 1} d\xi .
\end{aligned}
$$

Denote $\chi_\alpha := \alpha(1 - \alpha) B(1 - 2\alpha, \alpha + 1)$. Then from (2.9)-(2.11) it follows that

(2.12) $$ I_1 = I_2 + I_3 , $$

where

$$
\begin{aligned}
(2.13) \qquad I_2 &:= \chi_\alpha \int_u^{t_2} \int_0^u \xi^{-\alpha} (v - \xi)^{2\alpha - 1} d\xi dv \\
&= \frac{\chi_\alpha}{2\alpha} \int_0^u \xi^{-\alpha} \left[(t_2 - \xi)^{2\alpha} - (u - \xi)^{2\alpha} \right] d\xi \\
&= \frac{\chi_\alpha}{2\alpha} \left[\int_0^u \xi^{-\alpha} (t_2 - \xi)^{2\alpha} d\xi - u^{\alpha + 1} B(1 - \alpha, 2\alpha + 1) \right]
\end{aligned}
$$

and

$$I_3 := \int_0^u v^\alpha \int_0^1 \eta^{-2\alpha}(1-\eta)^\alpha \left(\frac{u}{v}-\eta\right)^\alpha d\eta dv$$
$$-\alpha u \int_0^u v^{\alpha-1} \int_0^1 \eta^{-2\alpha}(1-\eta)^\alpha \left(\frac{u}{v}-\eta\right)^{\alpha-1} d\eta dv$$
$$= \int_0^1 \eta^{-2\alpha}(1-\eta)^\alpha \int_0^u (u-v\eta)^\alpha dv d\eta$$
$$-\alpha u \int_0^1 \eta^{-2\alpha}(1-\eta)^\alpha \int_0^u (u-v\eta)^{\alpha-1} dv d\eta$$
$$= \frac{u^{\alpha+1}}{\alpha+1}\int_0^1 \eta^{-2\alpha-1}(1-\eta)^\alpha d\eta - \frac{u^{\alpha+1}}{\alpha+1}\int_0^1 \eta^{-2\alpha-1}(1-\eta)^{2\alpha+1} d\eta$$
$$(2.14) \quad +u^{\alpha+1}\int_0^1 \eta^{-2\alpha-1}(1-\eta)^{2\alpha} d\eta - u^{\alpha+1}\int_0^1 \eta^{-2\alpha-1}(1-\eta)^\alpha d\eta .$$

Substituting (2.13)-(2.14) into (2.12) we obtain

$$\frac{\partial I_1}{\partial u} = \frac{\chi_\alpha}{2\alpha}\left[u^{-\alpha}(t_2-u)^{2\alpha} - (\alpha+1)B(1-\alpha,2\alpha+1)u^\alpha\right]$$
$$+u^\alpha\Big[B(-2\alpha,\alpha+1) - B(-2\alpha,2\alpha+2)$$
$$+(\alpha+1)B(-2\alpha,2\alpha+1) - (\alpha+1)B(-2\alpha,\alpha+1)\Big].$$

Therefore

$$\int_0^{t_1} u^\alpha \frac{\partial I_1}{\partial u} = \frac{\chi_\alpha}{2\alpha}\int_0^{t_1}(t_2-u)^{2\alpha}du + \tilde\chi_\alpha \int_0^{t_1} u^{2\alpha}du$$
$$= \frac{\chi_\alpha}{2\alpha(2\alpha+1)}\left[t_2^{2\alpha+1} - (t_2-t_1)^{2\alpha+1}\right] + \frac{\tilde\chi_\alpha}{2\alpha+1}t_1^{2\alpha+1},$$

where

$$\tilde\chi_\alpha := B(-2\alpha,\alpha+1) - B(-2\alpha,2\alpha+2) + (\alpha+1)B(-2\alpha,2\alpha+1)$$
$$-(\alpha+1)B(-2\alpha,\alpha+1) - \frac{1}{2}(1+\alpha)(1-\alpha)B(1-\alpha,2\alpha+1)B(1-2\alpha,\alpha+1)$$
$$= \frac{\Gamma(1-2H)\Gamma(\frac{1}{2}+H)}{\Gamma\left(\frac{1}{2}-H\right)}.$$

Now it is straightforward to check that

$$\frac{\tilde\chi_\alpha}{2\alpha+1}\kappa_H^2 = \frac{1}{2}$$

and

$$\frac{\chi_\alpha}{2\alpha(2\alpha+1)}\kappa_H^2 = \frac{1}{2}.$$

Therefore

$$\mathbb{E}\left(Y_{t_1}Y_{t_2}\right) = \frac{1}{2}\left(t_1^{2H} + t_2^{2H} - (t_2-t_1)^{2H}\right).$$

This concludes the proof of the theorem.   □

REMARK 2.2. From the proof of this theorem we have

$$(2.15) \qquad \int_0^{t \wedge s} Z_H(t,r)Z_H(s,r)dr = \frac{1}{2}\left(t^{2H} + s^{2H} - |t-s|^{2H}\right).$$

This identity is useful in the following sections.

THEOREM 2.3. Let $\frac{1}{2} < H < 1$ and denote $K_H = (H - \frac{1}{2})\kappa_H$. Then

$$(2.16) \qquad B_t^H = \int_0^t \rho_H(t,s)B_s ds,$$

where
(2.17)
$$\rho_H(t,s) = K_H \left[(H - \frac{1}{2})s^{-H-\frac{1}{2}} \int_s^t u^{H-\frac{1}{2}}(u-s)^{H-\frac{3}{2}}du - s^{-H-\frac{1}{2}}t^{H+\frac{1}{2}}(t-s)^{H-\frac{3}{2}} \right].$$

Proof Notice that $\lim_{s \to 0} s^\beta B_s = 0$ a.s. if $\beta > -1/2$. By the integration by parts formula, we have

$$B_t^H = \int_0^t \frac{\partial}{\partial s} Z_H(t,s)B_s ds.$$

Making substitution $u = sv$ in Equation (2.5), we have

$$(2.18) \qquad Z_H(t,s) = (H - \frac{1}{2})\kappa_H s^{H-\frac{1}{2}} \int_1^{\frac{t}{s}} v^{H-\frac{1}{2}}(v-1)^{H-\frac{3}{2}}dv,$$

It is easy to check that $\frac{\partial}{\partial s} Z_H(t,s) = \rho_H(t,s)$, where $\rho_H(t,s)$ is given by (2.17). $\square$

THEOREM 2.4. Let $0 < H < 1/2$ and denote $K_H = (H - \frac{1}{2})\kappa_H$. Then

$$(2.19) \qquad B_t^H = \int_0^t \rho_H(t,s)(B_t - B_s)ds,$$

where
(2.20)
$$\rho_H(t,s) = K_H \left[-t^{H-\frac{1}{2}}s^{\frac{1}{2}-H}(t-s)^{H-\frac{3}{2}} - (H - \frac{1}{2})s^{-H-\frac{1}{2}} \int_s^t u^{H-\frac{3}{2}}(u-s)^{H-\frac{1}{2}}du \right].$$

Proof Notice that $\lim_{s \to 0}(t-s)^\beta(B_t-B_s) = 0$ a.s. if $\beta > -1/2$. Thus $Z_H(t,s)(B_t - B_s) \to 0$ when $s \uparrow t$ or $s \downarrow 0$. By the integration by parts formula, we have

$$B_t^H = \int_0^t \frac{\partial}{\partial s} Z_H(t,s)(B_t - B_s)ds,$$

where $Z_H(t,s)$ is given by (2.3). In (2.3), if we make substitution $u = sv$, then
(2.21)
$$Z_H(t,s) = \kappa_H \left[t^{H-\frac{1}{2}}s^{\frac{1}{2}-H}(t-s)^{H-\frac{1}{2}} - (H - \frac{1}{2})s^{H-\frac{1}{2}} \int_1^{\frac{t}{s}} v^{H-\frac{3}{2}}(v-1)^{H-\frac{1}{2}}dv \right].$$

Now it is straightforward to verify that $\frac{\partial}{\partial s} Z_H(t,s) = \rho_H(t,s)$, where $\rho_H(t,s)$ is given by (2.20). $\square$

REMARK 2.5. (i) In the following sections, we shall study transformations induced by the representations introduced in this section as well as their applications.

(ii) It is easy to check that $Z_H(t,s)$ is homogeneous in t and s of order $H - \frac{1}{2}$. Namely,

$$Z_H(\alpha t, \alpha s) = \alpha^{H-\frac{1}{2}} Z_H(t,s) \quad \forall \, \alpha, s, t \geq 0 \,.$$

(iii) We also define $Z_H(t,s) = 0$ for $0 < t < s < \infty$.

It is easy to prove

THEOREM 2.6. Let $K(t,s)$, $0 \leq s < t < \infty$ be well-defined and satisfy $\int_0^t |K(t,s)|^2 ds < \infty$ for all $t > 0$. Assume that

$$K(\alpha t, \alpha s) = \alpha^{\beta - \frac{1}{2}} K(t,s) \quad \forall \, \alpha, s, t \geq 0 \quad \text{for some} \quad \beta > 1/2 \,.$$

Then $X_t = \int_0^t K(t,s) dB_s$, $t \geq 0$ is self-similar with Hurst parameter β. Namely, $X_{\alpha t}$ and $\alpha^\beta X_t$, $0 \leq t < \infty$ has the same probability law.

REMARK 2.7. Let

$$\tilde{Z}_H(t,s) = \frac{1}{\Gamma(H + \frac{1}{2})} (t - s)^{H - \frac{1}{2}} F(H - \frac{1}{2}, \frac{1}{2} - H, H + \frac{1}{2}, 1 - \frac{t}{s}) \,,$$

where $F(a,b,c,z)$ is the hypergeometric function, then Decreusefond and Üstünel proved in [33] that $(\int_0^t \tilde{Z}_H(t,s) dB_s, t \geq 0)$ is a fractional Brownian motion (with a different constant). They use a technique of analytic extension. They also proved that when $1/2 < H < 1$

$$\tilde{Z}_H(t,s) = \frac{s^{\frac{1}{2} - H}}{\Gamma(H - \frac{1}{2})} \int_s^t u^{H - \frac{1}{2}} (u - s)^{H - \frac{3}{2}} du$$

which is (2.5) up to a constant.

CHAPTER 3

Induced Transformation I

In this section we consider the transformations induced by the representations (2.16) and (2.19). Namely we shall obtain some properties of the integral transformations with the kernel $\rho_H(t, s)$ corresponding to the representation (2.16) and (2.19). In the next section, applications to numerical simulation of fBm will be given. Recall $K_H = (H - \frac{1}{2})\sqrt{\dfrac{2H\Gamma(\frac{3}{2} - H)}{\Gamma(H + \frac{1}{2})\Gamma(2 - 2H)}}$. Let $T > 0$ be given and fixed throughout this paper. Define

$$(3.1) \qquad \mathbb{A}_H f(t) = \begin{cases} \int_0^t \rho_H(t, s) f(s) ds, & \text{when } \frac{1}{2} < H < 1 \\[2mm] \int_0^t \rho_H(t, s)\left[f(t) - f(s)\right] ds, & \text{when } 0 < H < \frac{1}{2}, \end{cases}$$

where ρ_H defined on $\{(t, s)\,;\, 0 < s < t < \infty\}$ is introduced in the precedent section:
(3.2)

$$\rho_H(t,s) = \begin{cases} K_H\left[(H - \frac{1}{2})s^{-H-\frac{1}{2}}\int_s^t u^{H-\frac{1}{2}}(u - s)^{H-\frac{3}{2}}du - s^{-H-\frac{1}{2}}t^{H+\frac{1}{2}}(t - s)^{H-\frac{3}{2}}\right] \\[2mm] \qquad\qquad when \quad 1/2 < H < 1 \\[4mm] K_H\left[-t^{H-\frac{1}{2}}s^{\frac{1}{2}-H}(t - s)^{H-\frac{3}{2}} - (H - \frac{1}{2})s^{-H-\frac{1}{2}}\int_s^t u^{H-\frac{3}{2}}(u - s)^{H-\frac{1}{2}}du\right] \\[2mm] \qquad\qquad when \quad 0 < H < 1/2. \end{cases}$$

Let $H_{0,\lambda,\alpha,p}$ be the set of continuous functions such that

$$\|f\|_{\lambda,\alpha,p} = \sup_{0 \le t \le T}\left(\int_0^t s^{-\lambda p}(t - s)^{-\alpha p}|f(s)|^p ds\right)^{1/p} < \infty.$$

It is easy to see that $H_{0,\lambda,\alpha,p}$ is a linear space and $\|\cdot\|_{\lambda,\alpha,p}$ is a norm. The completion of $H_{0,\lambda,\alpha,p}$ under this norm is denoted by $H_{\lambda,\alpha,p}$.

The case $\alpha = 0$ will be treated distinctly. Let $H_{0,\lambda,p}$ be the set of continuous functions such that

$$\|f\|_{\lambda,p} = \left(\int_0^T s^{-\lambda p}|f(s)|^p ds\right)^{1/p} < \infty.$$

The completion of $H_{0,\lambda,p}$ under this norm is denoted by $H_{\lambda,p}$.

THEOREM 3.1. *Let $\frac{1}{2} < H < 1$ and let*

(3.3)
$$\frac{1}{p} - H + \frac{1}{2} < \alpha < \frac{1}{p}, \quad \lambda > \frac{1}{p} + H - \frac{1}{2}.$$

Then the linear transformation $\mathbb{A}_H$ is bounded from $H_{\lambda,\alpha,p}$ to L_∞, where L_∞ is the Banach space of bounded measurable functions with the essential sup norm.

Proof It is easy to verify that when $\frac{1}{2} < H < 1$,

$$|\rho_H(t,s)| \le C_{H,T} s^{-H-\frac{1}{2}}(t-s)^{H-\frac{3}{2}}, \quad 0 < s < t \le T.$$

Therefore

$$\begin{aligned}
|\mathbb{A}_H f(t)| &\le C_{H,T} \int_0^t s^{-H-\frac{1}{2}}(t-s)^{H-\frac{3}{2}}|f(s)|ds \\
&\le C_{H,T} \left(\int_0^t s^{(-H-\frac{1}{2}+\lambda)q}(t-s)^{(H-\frac{3}{2}+\alpha)q}ds \right)^{1/q} \\
&\quad \left(\int_0^t s^{-\lambda p}|f(s)|^p(t-s)^{-\alpha p}ds \right)^{1/p}
\end{aligned}$$

The first integral is bounded if

$$(-H - \frac{1}{2} + \lambda)q > -1 \quad \text{and} \quad (H - \frac{3}{2} + \alpha)q > -1$$

which is implied by (3.3). Thus

$$\sup_{0 \le t \le T} |\mathbb{A}_H f(t)| \le C_{H,T,\lambda,\alpha,p}\|f\|_{\lambda,\alpha,p},$$

where $C_{H,T,\lambda,\alpha,p}$ denotes a constant depending on H, T, λ, α, p. □

If $\alpha = 0$, then we obtain

COROLLARY 3.2. *Let $1/2 < H < 1$. If $\lambda > \frac{1}{p} + H - \frac{1}{2}$, then $\mathbb{A}_H$ is bounded from $H_{\lambda,p}$ to L_∞.*

Now we study the transformation $\mathbb{A}_H$ for $0 < H < 1/2$. Let f be a continuous function on $[0, T]$. The Hölder norm (of exponent λ, $\lambda > 0$) is defined by

$$\|f\|_{H^\lambda} = \sup_{0 \le s < t \le T} \frac{|f(t) - f(s)|}{(t-s)^\lambda}.$$

Let H^λ denote the space of all Hölder continuous functions (of exponent λ), *i.e.*

$$H^\lambda = \{f : \|f\|_{H^\lambda} < \infty\}.$$

THEOREM 3.3. *Let $0 < H < 1/2$ and let $0 < T < \infty$. If $\lambda > \frac{1}{2} - H$, then $\mathbb{A}_H$ is a bounded operator from H^λ to L_∞. Namely, there is a constant $C_{\lambda,H,T}$ such that*

$$\sup_{0 \le t \le T} |\mathbb{A}_H f(t)| \le C_{\lambda,H,T}\|f\|_{H^\lambda}, \quad \forall\ f \in H^\lambda.$$

Proof From (2.20) it follows that when $0 < H < 1/2$,

$$\begin{aligned}
|\rho_H(t,s)| &\le C \left[t^{H-\frac{1}{2}}s^{\frac{1}{2}-H}(t-s)^{H-\frac{3}{2}} + s^{-H-\frac{1}{2}} \int_s^t u^{H-\frac{3}{2}}(u-s)^{H-\frac{1}{2}}du \right] \\
&=: \rho_{H,1}(t,s) + \rho_{H,2}(t,s),
\end{aligned}$$

where and in what follows C denotes a constant whose value varies from place to place. By the definition of $\mathbb{A}_H$, we obtain that for $0 < t < T$,

$$
\begin{aligned}
|\mathbb{A}_H f(t)| &\leq \int_0^t |\rho_H(t,s)||f(t) - f(s)|ds \\
&= \int_0^t |\rho_H(t,s)|(t-s)^\lambda \frac{|f(t) - f(s)|}{(t-s)^\lambda} ds \\
&\leq \int_0^t |\rho_H(t,s)|(t-s)^\lambda ds \|f\|_{H^\lambda} \\
&\leq (I_1(t) + I_2(t))\|f\|_{H^\lambda} \,,
\end{aligned}
$$

where $I_i(t) = \int_0^t \rho_{H,i}(t,s)(t-s)^\lambda ds$, $i = 1, 2$. First let us estimate $I_1(t)$.

$$
\begin{aligned}
I_1(t) &\leq \int_0^t \rho_{H,1}(t,s)(t-s)^\lambda ds \\
&\leq Ct^{H-\frac{1}{2}} \int_0^t s^{\frac{1}{2}-H}(t-s)^{H-\frac{3}{2}+\lambda} ds \\
&\leq Ct^{H-\frac{1}{2}}t^\lambda = Ct^{\lambda+H-\frac{1}{2}} \,.
\end{aligned}
$$

Since $\lambda > \frac{1}{2} - H$, we see that

$$
\sup_{0<t\leq T} I_1(t) < \infty \,.
$$

Now we estimate $I_2(t)$. It is obvious that

$$
\begin{aligned}
I_2(t) &\leq \int_0^t \rho_{H,2}(t,s)(t-s)^\lambda ds \\
&\leq Ct^\lambda \int_0^t s^{-H-\frac{1}{2}} \int_s^t u^{H-\frac{3}{2}}(u-s)^{H-\frac{1}{2}} du\, ds \\
&= Ct^\lambda \int_0^t u^{H-\frac{3}{2}} \int_0^u s^{-H-\frac{1}{2}}(u-s)^{H-\frac{1}{2}} ds\, du \\
&\leq Ct^\lambda \int_0^t u^{H-\frac{3}{2}} du = Ct^{\lambda+H-\frac{1}{2}} \,.
\end{aligned}
$$

Similar to the argument for $I_1(t)$, we have

$$
\sup_{0<t\leq T} I_2(t) < \infty \,.
$$

Hence

$$
\sup_{0<t\leq T} |\mathbb{A}_H f(t)| \leq C\|f\|_{H^\lambda} \,,
$$

proving the theorem.   □

REMARK 3.4. Many other properties of the operator $\mathbb{A}_H$ may be established by using the idea of fractional calculus. It will not be done in this article.

CHAPTER 4

Approximation

In this section we shall study the approximation of fBm by the polygonal approximation of sBm. The rate of convergence will be given. Assume that there are $\#\pi$ points in the uniform partition π of the given interval $[0, T]$. It is shown that when $0 < H < 1/2$ the rate of convergence is $(\#\pi)^{-H}$ and when $1/2 < H < 1$ the rate of convergence is $(\#\pi)^{H-1}$. Moreover, when $3/4 < H < 1$, it is shown that the division points can be re-distributed in such a way that there are *more points near* 0 *than near* T, and the rate of convergence is $(\#\pi)^{(H-1)(4H-2)}$. Note that when $H > 3/4$, $(1 - H)(4H - 2) > 1 - H$. This yields a higher convergence rate if $3/4 < H < 1$.

Let $\pi : 0 = t_0 < t_1 < \cdots < t_{n-1} < t_n = T$ be a partition of the interval $[0, T]$, where $t_k = kT/n$. Denote $|\pi| := \max_{0 \le j \le n-1}(t_{j+1} - t_j) = T/n$. Let B_t^π be the polygonal approximation of the sBm B_t, $0 \le t \le T$ associated with this partition π. That is

$$B_t^\pi = B_{t_j} + \frac{B_{t_{j+1}} - B_{t_j}}{t_{j+1} - t_j}(t - t_j), \quad t_j \le t \le t_{j+1}, \quad j = 0, 1, 2, \cdots, n - 1.$$

Define

$$B_t^{H,\pi} = \begin{cases} \int_0^t \rho_H(t, s) B_s^\pi ds & \text{if } H > 1/2 \\[2mm] \int_0^t \rho_H(t, s)\left(B_t^\pi - B_s^\pi\right) ds & \text{if } H < 1/2. \end{cases}$$

It is natural to guess that $B_t^{H,\pi} \to B_t^H$ as $\pi \to 0$. It is shown for example in [33] that

$$\lim_{|\pi| \to \infty} \mathbb{E}\left(\int_0^1 |B_t^{H,\pi} - B_t^H|^2 dt\right) = 0.$$

In this section we will discuss the convergence in sup norm. More precisely, we shall show $\lim_{|\pi| \to 0} \mathbb{E}\left(\sup_{0 \le t \le T} |B_t^{H,\pi} - B_t|\right) = 0$ and the rate of converge is also given. It should be pointed out that many other approximation schemes have been considered as well (see for instance, [104], [105]).

4.1. Rate of Convergence When $0 < H < 1/2$

Let $0 < H < 1/2$. We need the following lemma (in the case $\lambda = 0$ see [51] for more discussion).

LEMMA 4.1. *Let* $\pi : 0 = t_0 < t_1 < \cdots < t_{n-1} < t_n = T$ *be a partition of the interval* $[0, T]$, *where* $t_k = kT/n$. *Let* $0 < \lambda < 1/2$. *Then for any* $0 < \gamma < \frac{1}{2} - \lambda$, *there is a constant* $0 < C_{\gamma,\lambda,T} < \infty$, *independent of partition* π, *such that*

(4.1)
$$\mathbb{E}\left\{\sup_{0 < s < t < T} \frac{|B_t^\pi - B_s^\pi - (B_t - B_s)|}{(t - s)^\lambda}\right\} \le C_{\gamma,\lambda,T}\left(\frac{1}{n}\right)^\gamma.$$

Proof   Let $0 < p < \infty$ be given. Then

$$\mathbb{E}\left\{\sup_{0<s<t<T}\frac{|B_t^\pi - B_s^\pi - (B_t - B_s)|}{(t-s)^\lambda}\right\}$$

$$\leq \mathbb{E}\left\{\max\left(\max_{0\leq j<k-1\leq n-2}\sup_{t_j<s<t_{j+1},t_k<t<t_{k+1}}\frac{|B_t^\pi - B_s^\pi - (B_t - B_s)|}{(t-s)^\lambda},\right.\right.$$

$$\left.\left.\max_{0\leq k\leq n-2}\sup_{t_k<s<t<t_{k+2}}\frac{|B_t^\pi - B_s^\pi - (B_t - B_s)|}{(t-s)^\lambda}\right)\right\}$$

$$\leq C_p\left\{\sum_{0\leq j<k-1\leq n-2}\mathbb{E}\sup_{t_j<s<t_{j+1},t_k<t<t_{k+1}}\frac{|B_t^\pi - B_s^\pi - (B_t - B_s)|^p}{(t-s)^{\lambda p}}\right\}^{1/p}$$

$$+C_p\left\{\sum_{0\leq k\leq n-2}\mathbb{E}\sup_{t_k<s<t<t_{k+2}}\frac{|B_t^\pi - B_s^\pi - (B_t - B_s)|^p}{(t-s)^{\lambda p}}\right\}^{1/p}$$

$$=:\quad I_1 + I_2.$$

In the following C_p denotes a constant dependent on p, whose value may vary from place to place. Notice that if $0 \leq j < k-1 \leq n-2$, $t_j < s < t_{j+1}$, and $t_k < t < t_{k+1}$, then $(t-s) > 1/n$. Moreover,

$$\mathbb{E}\left(\sup_{t_j<s<t_{j+1}}|B_s^\pi - B_s|^p\right) \leq C_p(1/n)^{p/2}$$

and

$$\mathbb{E}\left(\sup_{t_k<t<t_{k+1}}|B_t^\pi - B_t|^p\right) \leq C_p(1/n)^{p/2}.$$

Thus I_1 can be estimated as follows

$$I_1 \leq C_p\left\{2^p\sum_{0\leq j<k-1\leq n-2}\left[\mathbb{E}\sup_{t_k<t<t_{k+1}}\frac{|B_t^\pi - B_t|^p}{(t-s)^{\lambda p}}+\mathbb{E}\sup_{t_j<s<t_{j+1}}\frac{|B_s^\pi - B_s|^p}{(t-s)^{\lambda p}}\right]\right\}^{1/p}$$

$$\leq C_p\left\{\sum_{0\leq j<k-1\leq n-2}(1/n)^{(\frac{1}{2}-\lambda)p}\right\}^{1/p}$$

$$\leq C_p\left\{(1/n)^{(\frac{1}{2}-\lambda)p-2}\right\}^{1/p} \leq C_p(1/n)^{\frac{1}{2}-\lambda-2/p}.$$

If $\gamma < \frac{1}{2} - \lambda$, then we can choose p such that $\gamma = \frac{1}{2} - \lambda - 2/p$. This implies that

$$I_1 \leq C_p(1/n)^\gamma.$$

Now we estimate I_2.

$$I_2 = C_p\left\{\sum_{0\leq k\leq n-2}\mathbb{E}\sup_{t_k<s<t<t_{k+2}}\frac{|B_t^\pi - B_s^\pi - (B_t - B_s)|^p}{(t-s)^{\lambda p}}\right\}^{1/p}$$

$$\leq C_p\left\{\sum_{0\leq k\leq n-2}(1/n)^{(\frac{1}{2}-\lambda)p}\right\}^{1/p} = C_p(1/n)^{\frac{1}{2}-\lambda-1/p}.$$

Similar to the argument for I_1, we have for any $\gamma < \frac{1}{2} - \lambda$, there is a C_p such that

$$I_2 \le C_p (1/n)^\gamma.$$

This proves the lemma. $\qquad\square$

THEOREM 4.2. *Let* $0 < H < 1/2$. *For any* $0 < \gamma < H$, *there is a constant* $C_{H,\gamma,T}$, *independent of partition* π, *such that*

$$(4.2) \qquad \mathbb{E} \sup_{0 \le t \le T} |B_t^{H,\pi} - B_t^H| \le C_{H,\gamma,T} |\pi|^\gamma.$$

Proof Write

$$B_t^{H,\pi} - B_t^H = \int_0^t \rho(t,s)(B_t^\pi - B_s^\pi - (B_t - B_s))ds.$$

Set

$$\Xi_T = \sup_{0 < s < t \le T} \frac{|B_t^\pi - B_s^\pi - (B_t - B_s)|}{(t-s)^\lambda}.$$

We obtain

$$
\begin{aligned}
\left| B_t^{H,\pi} - B_t^H \right| &\le \int_0^t |\rho_H(t,s)||B_t^\pi - B_s^\pi - (B_t - B_s)|ds \\
&\le \int_0^t |\rho_H(t,s)|(t-s)^\lambda \frac{|B_t^\pi - B_s^\pi - (B_t - B_s)|}{(t-s)^\lambda} ds \\
&\le C \int_0^t |\rho_H(t,s)|(t-s)^\lambda \Xi_T ds \\
&\le C \left[\int_0^t s^{\frac{1}{2}-H} t^{H-\frac{1}{2}}(t-s)^{H-\frac{3}{2}+\lambda} ds \right. \\
&\quad \left. + \int_0^t s^{-\frac{1}{2}-H} \int_s^t u^{H-\frac{3}{2}}(u-s)^{H-\frac{1}{2}}(t-s)^\lambda ds \right] \Xi_T \\
&\le C t^{H-\frac{1}{2}+\lambda} \Xi_T + C \int_0^t u^{H-\frac{3}{2}} \int_0^u s^{-\frac{1}{2}-H}(u-s)^{H-\frac{1}{2}}(t-s)^\lambda ds \Xi_T \\
&\le C t^{H-\frac{1}{2}+\lambda} \Xi_T.
\end{aligned}
$$

By lemma 4.1

$$\mathbb{E}(\Xi_T) \le C_{\gamma,\lambda,T} \left(\frac{1}{n} \right)^\gamma, \quad \forall\, 0 < \gamma < \frac{1}{2} - \lambda.$$

If $\lambda > \frac{1}{2} - H$, then

$$\mathbb{E} \sup_{0 \le t \le T} \left| B_t^{H,\pi} - B_t^H \right| \le C T^{H-\frac{1}{2}+\lambda} \mathbb{E}\,[\Xi_T] \le C_{\gamma,\lambda,T} \left(\frac{1}{n} \right)^\gamma$$

for all $0 < \gamma < \frac{1}{2} - \lambda$. This implies the theorem easily. $\qquad\square$

4.2. Rate of Convergence When $1/2 < H < 1$

Now let us consider the case $1/2 < H < 1$.

THEOREM 4.3. *Let $1/2 < H < 1$. For any $\gamma < 1 - H$, there is a constant $C_{H,\gamma,T}$, independent of partition π, such that*

$$(4.3) \qquad \mathbb{E} \sup_{0 \leq t \leq T} |B_t^{H,\pi} - B_t^H| \leq C_{H,\gamma,T} |\pi|^\gamma.$$

Proof Write

$$B_t^{H,\pi} - B_t^H = \int_0^t \rho(t,s)(B_s^\pi - B_s) ds.$$

From Corollary 3.2 it follows that for any $\lambda > \frac{1}{p} + H - \frac{1}{2}$, the following inequality holds

$$
\begin{aligned}
\mathbb{E} \sup_{0 \leq t \leq T} |B_t^{H,\pi} - B_t^H| &\leq C_{H,\lambda,p} \mathbb{E} \left(\int_0^T s^{-\lambda p} |B_s^\pi - B_s|^p ds \right)^{1/p} \\
&\leq C_{H,\lambda,p} \left(\int_0^T s^{-\lambda p} \mathbb{E} |B_s^\pi - B_s|^p ds \right)^{1/p} \\
&= C_{H,\lambda,p} \left(\int_0^{t_1} s^{-\lambda p} \mathbb{E} |B_s^\pi - B_s|^p ds + \int_{t_1}^T s^{-\lambda p} \mathbb{E} |B_s^\pi - B_s|^p ds \right)^{1/p} \\
&=: C_{H,\lambda,p} (I_1 + I_2)^{1/p}.
\end{aligned}
$$

Now when $t_k \leq s \leq t_{k+1}$,

$$(4.4) \qquad \mathbb{E} |B_s^\pi - B_s|^p \leq C_p (t_{k+1} - t_k)^{p/2}.$$

Thus

$$I_1 \leq C \int_0^{t_1} s^{-\lambda p + p/2} ds \leq C \left(\frac{1}{n} \right)^{\frac{p}{2} - \lambda p + 1}.$$

Now we are going to estimate I_2. If $\lambda p > 1$, then

$$\sum_{k=1}^\infty k^{-\lambda p} < \infty.$$

Therefore

$$
\begin{aligned}
I_2 &\leq \sum_{k=1}^{n-1} \int_{t_k}^{t_{k+1}} s^{-\lambda p} (s - t_k)^{\frac{p}{2}} ds \\
&\leq C \sum_{k=1}^{n-1} t_k^{-\lambda p} (t_{k+1} - t_k)^{p/2+1} \\
&\leq C \sum_{k=1}^\infty k^{-\lambda p} \left(\frac{1}{n} \right)^{p/2+1-\lambda p} \leq C \left(\frac{1}{n} \right)^{p/2+1-\lambda p}.
\end{aligned}
$$

Consequently,

$$\mathbb{E} \sup_{0 \leq t \leq T} |B_t^{H,\pi'} - B_t^H| \leq C \left(\frac{1}{n} \right)^{1/2+1/p-\lambda}.$$

For any $0 < \gamma < 1 - H$, it is possible to find λ and p such that

$$\gamma < \frac{1}{2} + \frac{1}{p} - \lambda \quad \text{and} \quad \lambda > \frac{1}{p} + H - \frac{1}{2}.$$

(which implies $\lambda p > 1$.) This proves the theorem. $\square$

The above theorem 4.3 can also be written as

$$(4.5) \qquad \mathbb{E} \sup_{0 \le t \le T} |B_t^{H,\pi} - B_t^H| \le C_{H,\gamma,T} |\#\pi|^{-\gamma},$$

where $0 < \gamma < 1 - H$ and $\#\pi$ denotes the number of division points used in the partition π. This is the quantity relevant to the complexity of the computation when one uses above scheme in the approximation of B_t^H.

4.3. Higher Order of Convergence When $3/4 < H < 1$

Since $\rho_H(t,s)$ is singular when $s = 0$, it is natural to wonder that a finer partition near 0 would presumably give fast rate of convergence. The following theorem illustrates that this is true when $3/4 < H < 1$. The rate of convergence can be improved if the length of intervals near 0 is smaller. In the following discussion of the rate of convergence it is more appropriate to use the number of division points of π (denoted by $\#\pi$) rather than the length of the partition (*i.e.* $|\pi| = \max_{0 \le k \le n-1}(t_{k+1} - t_k)$).

Let $0 < \beta < 1$ and denote by $n_\beta = [n^\beta]$ the integer part of n^β (the biggest integer which is less than or equal to n^β). Let

$$\pi': 0 = t_{1,0} < t_{1,1} < \cdots < t_{1,n-1} < t_{1,nn_\beta} = t_{n_\beta} < t_{n_\beta+1} < \cdots < t_{n-1} < t_n = T$$

be a partition of the interval $[0,T]$, where $t_j = jT/n$, $0 \le j \le n$, and $t_{1,k} = kT/n^2$, $k = 0, 1, \cdots, nn_\beta$. Denote $|\pi'| := \max_{0 \le j \le n-1}(t_{j+1} - t_j)$. Let $B_t^{\pi'}$ be the polygonal approximation of the sBm B_t, $0 \le t \le T$. Set

$$B_t^{H,\pi'} = \int_0^t \rho_H(t,s) B_s^{\pi'} ds.$$

Let $\#\pi'$ be the number of division points employed in the partition π'.

THEOREM 4.4. *For any* $0 < \gamma < (1-H)(4H-2)$, *there is a constant* $C_{H,\gamma,T}$, *independent of partition* π, *such that*

$$(4.6) \qquad \mathbb{E} \sup_{0 \le t \le T} |B_t^{H,\pi'} - B_t^H| \le C_{H,\gamma,T} |\#\pi'|^{-\gamma}.$$

Proof   First we have

$$B_t^{H,\pi'} - B_t^H = \int_0^t \rho(t,s)(B_s^{\pi'} - B_s) ds.$$

From Corollary 3.2 it follows that for any $\lambda > \frac{1}{p} + H - \frac{1}{2}$, the following inequality holds

$$\mathbb{E} \sup_{0 \le t \le T} |B_t^{H,\pi'} - B_t^H| \;\le\; C_{H,\lambda,p} \mathbb{E} \left(\int_0^T s^{-\lambda p} |B_s^{\pi'} - B_s|^p ds \right)^{1/p}$$

$$\le\; C_{H,\lambda,p} \left(\int_0^T s^{-\lambda p} \mathbb{E} |B_s^{\pi'} - B_s|^p ds \right)^{1/p}.$$

It is easy to see that when $t_{1,k} \le t \le t_{1,k+1}$,

$$(4.7) \qquad \mathbb{E} |B_s^{\pi'} - B_s|^p \le C_p (t_{1,k+1} - t_{1,k})^{p/2}$$

and when $t_k \leq t \leq t_{k+1}$,

$$\text{(4.8)} \qquad\qquad \mathbb{E}\,|B_s^{\pi'} - B_s|^p \leq C_p(t_{k+1} - t_k)^{p/2}\,.$$

Therefore

$$\text{(4.9)} \qquad\qquad \int_0^T s^{-\lambda p}\mathbb{E}\,|B_s^{\pi'} - B_s|^p ds \leq I_1 + I_2 + I_3\,,$$

where

$$I_1 = C\left(\int_0^{t_{1,1}} s^{-\lambda p}\mathbb{E}\,|B_s^{\pi'} - B_s|^p ds\right)^{1/p}\;;$$

$$I_2 = C\left(\int_{t_{1,1}}^{t_{n_\beta}} s^{-\lambda p}\mathbb{E}\,|B_s^{\pi'} - B_s|^p ds\right)^{1/p}\;;$$

and

$$I_3 = C\left(\int_{t_{n_\beta}}^{T} s^{-\lambda p}\mathbb{E}\,|B_s^{\pi'} - B_s|^p ds\right)^{1/p}\,.$$

Now we estimate I_1, I_2, and I_3 separately. Recall $t_{1,1} = \frac{T}{n^2}$. Then

$$
\begin{aligned}
I_1 &\leq C\left(\int_0^{t_{1,1}} s^{-\lambda p}s^{\frac{p}{2}}ds\right)^{1/p} \leq Ct_{1,1}^{\frac{1}{2}-\lambda+\frac{1}{p}} \\
&\leq C(1/n^2)^{\frac{1}{2}-\lambda+\frac{1}{p}} \leq C(1/n)^{1-2\lambda+\frac{2}{p}}\,.
\end{aligned}
$$

If $\lambda p > 1$, then

$$
\begin{aligned}
I_2 &\leq C\left(\sum_{k=1}^{nn_\beta-1}\int_{t_{1,k}}^{t_{1,k+1}} s^{-\lambda p}(s-t_{1,k})^{\frac{p}{2}}ds\right)^{1/p} \\
&\leq C\left(\sum_{k=1}^{nn_\beta-1} t_{1,k}^{-\lambda p}(t_{1,k+1}-t_{1,k})^{p/2+1}\right)^{1/p} \leq C\left[\sum_{k=1}^{nn_\beta-1} k^{-\lambda p}\left(\frac{1}{n^2}\right)^{p/2+1-\lambda p}\right]^{1/p} \\
&\leq C\left[\left(\frac{1}{n^2}\right)^{p/2+1-\lambda p}\right]^{1/p} \leq C\left(\frac{1}{n}\right)^{1+2/p-2\lambda}\,.
\end{aligned}
$$

I_3 is estimated as follows.

$$
\begin{aligned}
I_3 &\leq C\left(\sum_{k=n_\beta}^{n-1}\int_{t_k}^{t_{k+1}} s^{-\lambda p}(s-t_k)^{\frac{p}{2}}ds\right)^{1/p} \\
&\leq C\left(\sum_{k=n_\beta}^{n-1} t_k^{-\lambda p}(t_{k+1}-t_k)^{p/2+1}\right)^{1/p} \leq C\left[\sum_{k=n_\beta}^{n-1} k^{-\lambda p}\left(\frac{1}{n}\right)^{p/2+1-\lambda p}\right]^{1/p} \\
&\leq C\left[\left(\frac{1}{n}\right)^{p/2+1-\lambda p+\lambda\beta p-\beta}\right]^{1/p} \leq C\left(\frac{1}{n}\right)^{\frac{1}{2}-\lambda+\frac{1}{p}+\lambda\beta-\frac{\beta}{p}}\,.
\end{aligned}
$$

If we choose β, λ, and p such that

$$\text{(4.10)} \qquad\qquad \frac{1}{2} - \lambda + \frac{1}{p} + \lambda\beta - \frac{\beta}{p} = 1 + 2/p - 2\lambda\,,$$

then

$$\mathbb{E} \sup_{0 \le t \le T} |B_t^{H,\pi'} - B_t^H| \le C \left(\frac{1}{n} \right)^{1+2/p-2\lambda}.$$

If $\lambda = \frac{1}{p} + H - \frac{1}{2}$, then $1 + 2/p - 2\lambda = 2 - 2H$. Now the joint equation

$$\frac{1}{2} - \lambda + \frac{1}{p} + \lambda\beta - \frac{\beta}{p} = 1 + 2/p - 2\lambda \quad \text{and} \quad \lambda = \frac{1}{p} + H - \frac{1}{2},$$

yields $\beta = \frac{1-H}{H-\frac{1}{2}}$, which is greater than 0, and smaller than 1 when $H > \frac{3}{4}$. It is easy to argue that if $0 < \gamma < 2 - 2H$ and $H > 3/4$, then there is an λ, β and p such that

$$0 < \beta < 1, \quad \lambda > \frac{1}{p} + H - \frac{1}{2}, \quad \frac{1}{2} - \lambda + \frac{1}{p} + \lambda\beta - \frac{\beta}{p} = 1 + 2/p - 2\lambda,$$

and

$$1 + \frac{2}{p} - 2\lambda \ge \gamma.$$

Therefore

$$\mathbb{E} \sup_{0 \le t \le T} |B_t^{H,\pi'} - B_t^H| \le C \left(\frac{1}{n} \right)^\gamma.$$

Now the number of division points of the partition π' is $\#\pi'$ is of the order $n^{1+\beta}$. Hence,

$$(4.11) \qquad \mathbb{E} \sup_{0 \le t \le T} |B_t^{H,\pi'} - B_t^H| \le C \left(\frac{1}{\#\pi'} \right)^{\frac{\gamma}{1+\beta}}.$$

From $\beta = \frac{1-H}{H-\frac{1}{2}}$ we have $1 + \beta = \frac{1}{2H-1}$. Consequently,

$$(4.12) \qquad \frac{\gamma}{1+\beta} = \frac{2-2H}{1+\beta} = (2-2H)(2H-1) = (1-H)(4H-2)$$

which is greater than $1 - H$ if $H > 3/4$. Combining (4.11) and (4.12), we obtain the theorem. $\qquad \square$

4.4. Best Approximation

Let the standard Brownian motion be given at points $0 = t_0 < t_1 < \cdots < t_n = T$, namely the random variables $B_{t_1}, \cdots, B_{t_n}$ are known ($B_0 = 0$). If we use a function of $B_{t_1}, \cdots, B_{t_n}$, i.e. $g(t, B_{t_1}, \cdots, B_{t_n})$ to approximate B_t^H, it is interesting to know the best approximation in the sense that minimizes

$$(4.13) \qquad \mathbb{E} |g(t, B_{t_1}, \cdots, B_{t_n}) - B_t^H|^2$$

among all possible (measurable) function g and for all $t \in [0, T]$, where B_t^H is the fractional Brownian motion defined by (2.3). First we know that for any fixed t, the function g which minimizes $\mathbb{E} |g(t, B_{t_1}, \cdots, B_{t_n}) - B_t^H|^2$ is the conditional expectation of B_t^H given $B_{t_1}, \cdots, B_{t_n}$:

$$g^*(t) := g^*(t, B_{t_1}, \cdots, B_{t_n}) = \mathbb{E} \left[B_t^H | B_{t_1}, \cdots, B_{t_n} \right].$$

In the remaining part of this section, we shall show that $g^*(t) = B_t^{H,\pi}$. Denote $X_k = B_{t_k} - B_{t_{k-1}}$, $k = 1, 2, \cdots, n$. Then the σ-algebras generated by $B_{t_1}, \cdots, B_{t_n}$ and $X_1, \cdots, X_n$ are the same. Let

$$A_{ij} = \mathbb{E} (X_i X_j) = \delta_{ij} (t_{j+1} - t_j)$$

and let $\theta_k(t) = \mathbb{E}\left(B_t^H X_k\right)$. Let D_{ij} be the element of the inverse matrix of $A = (A_{ij})_{1 \le i,j \le n}$. Then from the conditional expectation properties of Gaussian random variables we have

$$g^*(t) = \sum_{i,j=1}^n D_{ij} \theta_i(t) X_j\,.$$

It is easy to see that $D_{ij} = \delta_{ij}(t_{j+1} - t_j)^{-1}$, where δ_{ij} is the Kronecker symbol, $i.e.$ $\delta_{ij} = 1$ if $i = j$ and $\delta_{ij} = 0$ if $i \ne j$. Moreover,

$$\theta_k(t) = \int_{t_{k-1}}^{t_k \wedge t} Z_H(t,s) ds \quad \text{if} \quad t_{k-1} \le t \le t_k\,.$$

Thus when $t_n \le t < t_{n+1}$

$$
\begin{aligned}
g^*(t) &= \sum_{k=1}^n \int_{t_{k-1}}^{t_k \wedge t} Z_H(t,s) ds \frac{B_{t_k} - B_{t_{k-1}}}{t_{k+1} - t_k} + \int_{t_n}^t Z_H(t,s) ds \frac{B_{t_k} - B_{t_{k-1}}}{t_{k+1} - t_k} \\
&= \int_0^t Z_H(t,s) \dot{B}_t^\pi ds\,,
\end{aligned}
$$

where B_t^π is the polygonal approximation of B_t. Now an integration by parts formula yields

$$g^*(t) = \begin{cases} \int_0^t \rho_H(t,s) B_t^\pi ds & \text{when} \quad 1/2 < H < 1 \\ \int_0^t \rho_H(t,s)\left(B_t^\pi - B_s^\pi\right) ds & \text{when} \quad 0 < H < 1/2\,. \end{cases}$$

Thus $g^*(t) = B_t^{H,\pi}$, $0 \le t \le T$. This means that $B^{H,\pi}$ is the best approximation in the mean square error sense.

REMARK 4.5. In a similar way we can prove that for any $1 \le p < \infty$ and for any γ satisfying

$$
(4.14) \qquad \begin{cases} 0 < \gamma < H & \text{if} \ \ 0 < H < 1/2 \\ 0 < \gamma < 1 - H & \text{if} \ \ 1/2 < H < 3/4 \\ 0 < \gamma < (1-H)(4H-2) & \text{if} \ \ 3/4 < H < 1 \end{cases}
$$

there is a positive $C_{H,T,\gamma,p}$, independent of partition π, such that

$$(4.15) \qquad \mathbb{E} \sup_{0 \le t \le T} |B_t^{H,\pi} - B_t^H|^p \le C_{H,T,\gamma,p} |\pi^*|^{-\gamma p}\,.$$

CHAPTER 5

Induced Transformation II

5.1. Operators Associated With $Z_H(t,s)$

In this section we study the transformation induced by the representation (2.3). Similar transformations appeared in [**33**], [**47**], [**48**], [**58**], [**90**]. We shall carry out a more detailed study. Its applications to Girsanov transformation, conditioning, continuity, and stochastic control will be given in the sections to follow.

If we formally differentiate (2.3) with respect to t, then we have the following heuristic equality

$$(5.1) \qquad \dot{B}_t^H = \frac{d}{dt} B_t^H = \frac{d}{dt} \int_0^t Z_H(t,s) dB_s = \frac{d}{dt} \int_0^t Z_H(t,s) \dot{B}_s ds.$$

Motivated by this equation we introduce the following integro-differential transformation:

$$(5.2) \qquad \mathbb{I}_{H,T} f(t) = \frac{d}{dt} \int_0^t Z_H(t,s) f(s) ds, \quad 0 < t < T,$$

where $Z_H(t,s)$ is given by (2.4) and (2.5), *i.e.*
(5.3)
$$Z_H(t,s) = \begin{cases} \kappa_H \left[\left(\frac{t}{s}\right)^{H-\frac{1}{2}} (t-s)^{H-\frac{1}{2}} - (H-\frac{1}{2}) s^{\frac{1}{2}-H} \int_s^t u^{H-\frac{3}{2}} (u-s)^{H-\frac{1}{2}} du \right] \\ \qquad \text{if } 0 < H < 1/2 \\ \\ (H-\frac{1}{2}) \kappa_H s^{\frac{1}{2}-H} \int_s^t u^{H-\frac{1}{2}} (u-s)^{H-\frac{3}{2}} du \\ \qquad \text{if } 1/2 < H < 1. \end{cases}$$

In the case of no ambiguity, we also use $\mathbb{I}_H = \mathbb{I}_{H,T}$.

Equation (5.1) suggests that formally $\mathbb{I}_H$ is a transformation which transforms the white noise (the derivative of sBm) to the fractional noise (the derivative of fBm):

$$\mathbb{I}_H \dot{B} = \dot{B}^H.$$

This heuristic link of $\mathbb{I}_H$ between sBm and fBm indicates that this transformation should play an important role in the theory of fBm, which will be elaborated in the following sections.

First let us discuss the domain and image of the above transformation. Certainly one can define the domain and image in many different ways. We restrict our discussion in the applications of $\mathbb{I}_H$ to the study of stochastic calculus of fBm.

Let $0 < T < \infty$. We shall consider functions over the interval $[0,T]$. Most of the theorems below are also valid on the interval $[0,\infty)$.

27

THEOREM 5.1. *(i) If $1/2 < H < 1$ and if f is Borel measurable and bounded over the interval $[0, T]$, then $\mathbb{I}_H$ is well-defined and*

$$(5.4) \qquad \mathbb{I}_H f(t) = \int_0^t \frac{\partial}{\partial t} Z_H(t, s) f(s) ds\,, \quad 0 \le t \le T\,.$$

(ii) If $0 < H < 1/2$ and if f is continuously differentiable over interval $[0, T]$, then $\mathbb{I}_H$ is well-defined.

Proof Notice that when $1/2 < H < 1$, $Z_H(t, s)$ is differentiable with respect to t, $Z_H(t, t) = 0$, and

$$(5.5) \qquad \frac{\partial}{\partial t} Z_H(t, s) = \kappa_H \left(H - \frac{1}{2} \right) s^{\frac{1}{2} - H} t^{H - \frac{1}{2}} (t - s)^{H - \frac{3}{2}}\,.$$

If f is Borel measurable and bounded, then

$$\int_0^t |\frac{\partial}{\partial t} Z_H(t, s) f(s)| ds < \infty\,.$$

So when $1/2 < H < 1$, $\mathbb{I}_H f$ is well-defined for all Borel measurable and bounded function f and we have

$$\mathbb{I}_H f(t) = \int_0^t \frac{\partial}{\partial t} Z_H(t, s) f(s) ds\,.$$

Now let $0 < H < 1/2$. If f is bounded and Borel measurable, then $\int_0^t Z_H(t, s) f(s) ds$ is well-defined since

$$|Z_H(t, s)| \le C s^{\frac{1}{2} - H} t^{H - \frac{1}{2}} (t - s)^{H - \frac{1}{2}}\,.$$

Assume now f is continuously differentiable. Denote

$$\begin{aligned} g(t) \quad &:= \quad \int_0^t Z_H(t, s) f(s) ds \\ &= \quad \kappa_H t^{H - \frac{1}{2}} \int_0^t s^{\frac{1}{2} - H} (t - s)^{H - \frac{1}{2}} f(s) ds \\ &\quad - \kappa_H \left(H - \frac{1}{2} \right) \int_0^t s^{\frac{1}{2} - H} \int_s^t u^{H - \frac{3}{2}} (u - s)^{H - \frac{1}{2}} du f(s) ds \\ &=: \quad I_1(t) + I_2(t)\,. \end{aligned}$$

It is obvious that $I_2(t)$ is differentiable with respect to t. Making substitution $s = t - r$ yields

$$I_1(t) = \kappa_H t^{H - \frac{1}{2}} \int_0^t (t - r)^{\frac{1}{2} - H} r^{H - \frac{1}{2}} f(t - r) dr\,.$$

Differentiating $I_1(t)$ with respect to t, we obtain

$$\begin{aligned} I_1'(t) \quad = \quad &\kappa_H \left(\frac{1}{2} - H \right) t^{H - \frac{1}{2}} \int_0^t (t - r)^{-\frac{1}{2} - H} r^{H - \frac{1}{2}} f(t - r) dr \\ &+ \kappa_H t^{H - \frac{1}{2}} \int_0^t (t - r)^{\frac{1}{2} - H} r^{H - \frac{1}{2}} f'(t - r) dr \\ &+ \kappa_H \left(H - \frac{1}{2} \right) t^{H - \frac{3}{2}} \int_0^t (t - r)^{\frac{1}{2} - H} r^{H - \frac{1}{2}} f(t - r) dr\,. \end{aligned}$$

It is easy to verify that all of above integrals are well-defined. Thus $I_1(t)$ is also differentiable with respect to t.   □

5.2. Inverse Operator of $\mathbb{I}_{H,T}$

Denote

$$\kappa_1 = \frac{1}{2H\Gamma(H+\frac{1}{2})\Gamma(\frac{3}{2}-H)}\,.$$

Introduce a transformation $\mathbb{B}_{H,T}$ as follows.

$$(5.6) \qquad \mathbb{B}_{H,T}f(r) = \frac{d}{dr}\int_0^r \eta_H(r,t)f(t)dt\,, \quad 0 \le r \le T\,,$$

where
(5.7)

$$\eta_H(r,t) = \begin{cases} \frac{2H\kappa_1}{\kappa_H}\left[r^{H-\frac{1}{2}}t^{\frac{1}{2}-H}(r-t)^{\frac{1}{2}-H} - (H-\frac{1}{2})t^{\frac{1}{2}-H}\int_t^r v^{H-\frac{3}{2}}(v-t)^{\frac{1}{2}-H}dv\right] \\ \qquad\qquad \text{if } 1/2 < H < 1 \\[2mm] \frac{H(1-2H)\kappa_1}{\kappa_H}t^{\frac{1}{2}-H}\int_t^r v^{H-\frac{1}{2}}(v-t)^{-\frac{1}{2}-H}dv \\ \qquad\qquad \text{if } 0 < H < 1/2\,. \end{cases}$$

When there is no ambiguity we denote $\mathbb{B} = \mathbb{B}_H = \mathbb{B}_{H,T}$. It is easy to see that when $0 < H < 1/2$, we have

$$(5.8) \qquad \mathbb{B}_{H,T}f(r) = \frac{H(1-2H)\kappa_1}{\kappa_H}r^{H-\frac{1}{2}}\int_0^r t^{\frac{1}{2}-H}(r-t)^{-\frac{1}{2}-H}f(t)dt\,.$$

EXAMPLE 5.2. Let $0 < H < 1/2$. If $f = 1$, then

$$\begin{aligned} [\mathbb{B}_{H,T}1]\,(t) &= \frac{H(1-2H)^2\kappa_1}{\kappa_H}t^{H-\frac{1}{2}}\int_0^t s^{\frac{1}{2}-H}(t-s)^{-\frac{1}{2}-H}ds \\ (5.9) \qquad &= \frac{H(1-2H)^2\kappa_1}{\kappa_H}B(\frac{3}{2}-H,\frac{1}{2}-H)t^{\frac{1}{2}-H}\,. \end{aligned}$$

In the following four subsections, we shall show that $\mathbb{B}_{H,T}$ is the inverse of $\mathbb{I}_{H,T}$.

5.3. $\mathbb{B}_{H,T}\mathbb{I}_{H,T}$ when $1/2 < H < 1$

THEOREM 5.3. *Let $\frac{1}{2} < H < 1$ and let f be continuous on $[0,T]$. Denote*

$$g(t) = \mathbb{I}_H f(t) = \frac{d}{dt}\int_0^t Z_H(t,s)f(s)ds\,, \quad 0 \le t \le T\,.$$

Then g is well-defined. Moreover, $\mathbb{B}_{H,T}g$ is well-defined and

$$(5.10) \qquad \mathbb{B}_{H,T}g(r) = \frac{d}{dr}\int_0^r \eta_H(r,t)g(t)dt = f(r)\,, \quad 0 \le r \le T\,.$$

REMARK 5.4. This theorem states that formally if $1/2 < H < 1$, then $\mathbb{B}_{H,T}\mathbb{I}_{H,T} = I$, where I is the identity operator.

Proof From the Fubini lemma it follows that

$$
\begin{aligned}
\mathbb{B}_{H,T}g(r) &= \frac{d}{dr}\int_0^r \eta_H(r,t)g(t)dt \\
&= \frac{d}{dr}\int_0^r \eta_H(r,t)\int_0^t \frac{\partial}{\partial t}Z_H(t,s)f(s)\,ds\,dt \\
&= \frac{d}{dr}\int_0^r \int_s^r \eta_H(r,t)\frac{\partial}{\partial t}Z_H(t,s)\,dt\,f(s)\,ds\,.
\end{aligned}
$$

Hence it suffices to show that

(5.11)
$$
\int_s^r \eta_H(r,t)\frac{\partial}{\partial t}Z_H(t,s)dt = 1, \quad 0 \le s < r < \infty.
$$

Using (5.5) we have

$$
\begin{aligned}
&\int_s^r \eta_H(r,t)\frac{\partial}{\partial t}Z_H(t,s)dt \\
&= 2H\kappa_1\left(H-\frac{1}{2}\right)\int_s^r \Bigg[r^{H-\frac{1}{2}}t^{\frac{1}{2}-H}(r-t)^{\frac{1}{2}-H}s^{\frac{1}{2}-H}t^{H-\frac{1}{2}}(t-s)^{H-\frac{3}{2}} \\
&\qquad -\left(H-\frac{1}{2}\right)s^{\frac{1}{2}-H}t^{H-\frac{1}{2}}(t-s)^{H-\frac{3}{2}}\int_t^r v^{H-\frac{3}{2}}t^{\frac{1}{2}-H}(v-t)^{\frac{1}{2}-H}dv \Bigg]dt \\
&= 2H\kappa_1\left(H-\frac{1}{2}\right)r^{H-\frac{1}{2}}s^{\frac{1}{2}-H}\int_s^r (t-s)^{H-\frac{3}{2}}(r-t)^{\frac{1}{2}-H}dt \\
&\qquad -2H\kappa_1\left(H-\frac{1}{2}\right)^2 s^{\frac{1}{2}-H}\int_s^r v^{H-\frac{3}{2}}\int_s^v (t-s)^{H-\frac{3}{2}}(v-t)^{\frac{1}{2}-H}dt\,dv \\
&= 2H\kappa_1\left(H-\frac{1}{2}\right)B\left(H-\frac{1}{2},\frac{3}{2}-H\right)r^{H-\frac{1}{2}}s^{\frac{1}{2}-H} \\
&\qquad -2H\kappa_1\left(H-\frac{1}{2}\right)^2 B\left(H-\frac{1}{2},\frac{3}{2}-H\right)s^{\frac{1}{2}-H}\int_s^r v^{H-\frac{3}{2}}dv \\
&= 2H\kappa_1\left(H-\frac{1}{2}\right)B\left(H-\frac{1}{2},\frac{3}{2}-H\right) = 1\,,
\end{aligned}
$$

proving the theorem. $\qquad\qquad\qquad\qquad\qquad\qquad\qquad\qquad\qquad\qquad\square$

5.4. $\mathbb{T}_{H,T}\mathbb{B}_{H,T}$ for $1/2 < H < 1$

THEOREM 5.5. *Let $H > \frac{1}{2}$ and let f be differentiable on $[0,T]$ and $f(0) = 0$. Then*

$$
g(t) = \mathbb{B}_{H,T}f(t) = \frac{d}{dt}\int_0^t \eta_H(t,s)f(s)ds, \quad 0 \le t \le T,
$$

is well-defined and $\mathbb{T}_H g$ is also well-defined. Moreover,

(5.12)
$$
\mathbb{T}_H g(t) = \frac{d}{dr}\int_0^r Z_H(r,t)g(t)dt = f(r), \quad 0 \le t \le T.
$$

Proof First it is easy to see that if f is differentiable, then $\int_0^t \eta_H(t,s)f(s)ds$ is well-defined. Now we show that it is differentiable. By definition of η_H and the substitution $s = tu$, we have

$$\int_0^t \eta_H(t,s)f(s)ds$$

$$= \quad C_{H,1}t^{H-\frac{1}{2}}\int_0^t s^{\frac{1}{2}-H}(t-s)^{\frac{1}{2}-H}f(s)ds$$

$$+ C_{H,2}\int_0^t s^{\frac{1}{2}-H}\int_s^t v^{H-\frac{3}{2}}(v-s)^{\frac{1}{2}-H}dvf(s)ds$$

$$= \quad C_{H,1}t^{\frac{3}{2}-H}\int_0^1 (1-u)^{\frac{1}{2}-H}u^{\frac{1}{2}-H}f(tu)du$$

$$+ C_{H,2}\int_0^t s^{\frac{1}{2}-H}\int_s^t v^{H-\frac{3}{2}}(v-s)^{\frac{1}{2}-H}dvf(s)ds$$

$$=: \quad I_1(t) + I_2(t),$$

where $C_{H,1}$ and $C_{H,2}$ are two constants. Differentiating $I_1(t)$ with respect to t, we obtain

$$I_1'(t) \quad = \quad C_{H,1}(\frac{3}{2}-H)t^{\frac{1}{2}-H}\int_0^1 (1-u)^{\frac{1}{2}-H}u^{\frac{1}{2}-H}f(tu)du$$

$$+ C_{H,1}t^{\frac{3}{2}-H}\int_0^1 (1-u)^{\frac{1}{2}-H}u^{\frac{3}{2}-H}f'(tu)du$$

From the fact that f is continuously differentiable and $f(0) = 0$ it follows that there is a $C_{H,f}$ with $\sup_{0\leq t\leq T}|C_{H,f}| < \infty$ such that $|f(tu)| \leq C_{H,f}tu$. Thus

$$|I_1'(t)| \leq C_{H,f}t^{\frac{3}{2}-H}.$$

Now we consider $I_2(t)$. We have

$$I_2'(t) \quad = \quad C_{H,2}t^{H-\frac{3}{2}}\int_0^t s^{\frac{1}{2}-H}(t-s)^{\frac{1}{2}-H}dsf(s)ds.$$

Hence

$$|I_2'(t)| \quad \leq \quad C_{H,f}t^{H-\frac{3}{2}}\int_0^t s^{\frac{1}{2}-H}(t-s)^{\frac{1}{2}-H}|f(s)|ds$$

$$\leq \quad C_{H,f}t^{H-\frac{3}{2}}\int_0^t s^{\frac{1}{2}-H}(t-s)^{\frac{1}{2}-H}sds$$

$$\leq \quad C_{H,f}t^{\frac{3}{2}-H}C_{H,t}t^{3-2H} = C_{H,f}t^{\frac{3}{2}-H}.$$

Therefore

(5.13) $$|g(t)| \leq C_{H,t}t^{\frac{3}{2}-H}.$$

From the definition of $\mathbb{I}_H$ and $\mathbb{B}_{H,T}$, it follows

$$\mathbb{I}_Hg(r) \quad = \quad \frac{d}{dr}\int_0^r Z_H(r,t)g(t)dt$$

$$= \quad \frac{d}{dr}\int_0^r Z_H(r,t)\frac{d}{dt}\int_0^t \eta_H(t,s)f(s)dsdt.$$

Notice that $\lim_{t\to 0} Z_H(r,t)g(t) = 0$. Application of integration by parts formula yields

$$\mathbb{T}_H g(r) = -\frac{d}{dr} \int_0^r \frac{\partial}{\partial t} Z_H(r,t) \int_0^t \eta_H(t,s) f(s) ds dt$$

$$= -\frac{d}{dr} \int_0^r \left\{ \int_s^r \frac{\partial}{\partial t} Z_H(r,t) \eta_H(t,s) dt \right\} f(s) ds\,.$$

It suffices to show that

$$\int_s^r \frac{\partial}{\partial t} Z_H(r,t) \eta_H(t,s) dt = -1\,.$$

From

$$\frac{\partial}{\partial t} Z_H(r,t) = \kappa_H \left(H - \frac{1}{2} \right) \Bigg[(H - \frac{1}{2}) t^{-H-\frac{1}{2}}$$

$$\int_t^r u^{H-\frac{1}{2}} (u-t)^{H-\frac{3}{2}} du - t^{-H-\frac{1}{2}} r^{H+\frac{1}{2}} (r-t)^{H-\frac{3}{2}} \Bigg],$$

it follows that

$$\int_s^r \frac{\partial}{\partial t} Z_H(r,t) \eta_H(t,s) dt = 2H\kappa_1 (H - \frac{1}{2}) \left[I_1 + I_2 + I_3 \right],$$

where I_1, I_2, and I_3 are defined and computed as follows.

$$I_1 := \left(H - \frac{1}{2} \right) \int_s^r t^{-H-\frac{1}{2}} \int_t^r u^{H-\frac{1}{2}} (u-t)^{H-\frac{3}{2}} du\, s^{\frac{1}{2}-H} t^{H-\frac{1}{2}} (t-s)^{\frac{1}{2}-H} dt$$

$$= \left(H - \frac{1}{2} \right) s^{\frac{1}{2}-H} \int_s^r u^{H-\frac{1}{2}} \int_s^u t^{-1} (t-s)^{\frac{1}{2}-H} (u-t)^{H-\frac{3}{2}} dt du\,.$$

From Lemma 14.4 with $\mu = \frac{3}{2} - H$, $\nu = H - \frac{1}{2}$ it follows that

$$I_1 = \left(H - \frac{1}{2} \right) s^{\frac{1}{2}-H} \int_s^r u^{H-\frac{1}{2}} B(\frac{3}{2} - H, H - \frac{1}{2}) s^{\frac{1}{2}-H} u^{H-\frac{3}{2}} du$$

$$(5.14) \qquad = \frac{1}{2} B(\frac{3}{2} - H, H - \frac{1}{2}) s^{1-2H} \left[r^{2H-1} - s^{2H-1} \right].$$

I_2 is given by

$$I_2 := -r^{H+\frac{1}{2}} \int_s^r t^{-H-\frac{1}{2}} (r-t)^{H-\frac{3}{2}} t^{H-\frac{1}{2}} s^{\frac{1}{2}-H} (t-s)^{\frac{1}{2}-H} dt\,.$$

Lemma 14.4 yields

$$(5.15) \qquad I_2 = -s^{1-2H} r^{2H-1} B(\frac{3}{2} - H, H - \frac{1}{2})\,.$$

Finally,

$$I_3 := -\frac{1}{\kappa_H} \int_s^r \frac{\partial}{\partial t} Z_H(r,t) \int_s^t v^{H-\frac{3}{2}} (v-s)^{\frac{1}{2}-H} s^{\frac{1}{2}-H} dv dt\,.$$

An integration by parts yields

$$
\begin{aligned}
I_3 &= \frac{1}{\kappa_H} \int_s^r Z_H(r,t) t^{H-\frac{3}{2}} (t-s)^{\frac{1}{2}-H} s^{\frac{1}{2}-H} dt \\
&= \frac{1}{\kappa_H} \int_s^r t^{\frac{1}{2}-H} \int_t^r u^{H-\frac{1}{2}} (u-t)^{H-\frac{3}{2}} du\, t^{H-\frac{3}{2}} (t-s)^{\frac{1}{2}-H} s^{\frac{1}{2}-H} dt \\
&= (H-\frac{1}{2}) s^{\frac{1}{2}-H} \int_s^r u^{H-\frac{1}{2}} \int_s^u t^{-1}(u-t)^{H-\frac{3}{2}}(t-s)^{\frac{1}{2}-H} dt\, du \,.
\end{aligned}
$$

Lemma 14.4 with $\mu = \frac{3}{2} - H$, $\nu = H - \frac{1}{2}$ yields

$$
\begin{aligned}
I_3 &= (H-\frac{1}{2}) B(\frac{3}{2}-H, H-\frac{1}{2}) s^{\frac{1}{2}-H} \int_s^r u^{H-\frac{1}{2}} u^{H-\frac{3}{2}} s^{\frac{1}{2}-H} du \\
&= \frac{1}{2} B(\frac{3}{2}-H, H-\frac{1}{2}) s^{1-2H} \left[r^{2H-1} - s^{2H-1} \right] \,.
\end{aligned}
$$
(5.16)

Combining (5.14)-(5.16) we obtain

$$
\int_s^r \frac{\partial}{\partial t} Z_H(r,t) \eta_H(t,s) dt = -2H\kappa_1 (H-\frac{1}{2}) B(\frac{3}{2}-H, H-\frac{1}{2}) = -1 \,,
$$

proving the theorem. $\square$

    This theorem states that $\mathbb{I}_H \mathbb{B}_{H,T} = I$ for $1/2 < H < 1$. Now we turn to consider similar identities for $0 < H < 1/2$.

5.5. $\mathbb{B}_{H,T}\mathbb{I}_{H,T}$ for $0 < H < 1/2$

Now we discuss $\mathbb{B}_{H,T}\mathbb{I}_H$ for $0 < H < 1/2$.

THEOREM 5.6. *Let $0 < H < \frac{1}{2}$ and let f be differentiable on $[0,T]$ with $f(0) = 0$. Then*

$$
g(t) = \mathbb{I}_H f(t) = \frac{d}{dt} \int_0^t Z_H(t,s) f(s) ds \,, \quad 0 \le t \le T
$$

is well-defined and $\mathbb{B}_{H,T} g$ is also well-defined. Moreover,

$$
\mathbb{B}_{H,T} g(r) = \frac{d}{dr} \int_0^r \eta_H(r,t) g(t) dt = f(r) \,, \quad 0 \le r \le T \,.
$$
(5.17)

Proof From the definition of η_H it follows easily that

$$
\begin{aligned}
\frac{\partial}{\partial t} \eta_H(r,t) &= \frac{H(1-2H)\kappa_1}{\kappa_H} \frac{\partial}{\partial t} \left\{ t^{\frac{1}{2}-H} \int_t^r v^{H-\frac{1}{2}}(v-t)^{-\frac{1}{2}-H} dv \right\} \\
&\quad - \frac{H(1-2H)\kappa_1}{\kappa_H} \frac{\partial}{\partial t} \left\{ t^{\frac{1}{2}-H} \int_1^{\frac{r}{t}} u^{H-\frac{1}{2}}(u-1)^{-\frac{1}{2}-H} du \right\} \\
&= \frac{H(1-2H)\kappa_1}{\kappa_H} t^{-\frac{1}{2}-H} \left[-r^{H+\frac{1}{2}}(r-t)^{-\frac{1}{2}-H} \right. \\
&\quad \left. + \left(\frac{1}{2}-H\right) \int_t^r v^{H-\frac{1}{2}}(v-t)^{-\frac{1}{2}-H} dv \right] .
\end{aligned}
$$
(5.18)

An integration by parts yields

$$
\begin{aligned}
\mathbb{B}_{H,T}g(r) &= \frac{d}{dr}\int_0^r \eta_H(r,t)\frac{d}{dt}\int_0^t Z_H(t,s)f(s)ds\,dt \\
&= -\frac{d}{dr}\int_0^r \frac{\partial}{\partial t}\eta_H(r,t)\int_0^t Z_H(t,s)f(s)ds\,dt \\
&= -\frac{d}{dr}\int_0^r \left[\int_s^r \frac{\partial}{\partial t}\eta_H(r,t)Z_H(t,s)dt\right]f(s)ds\,.
\end{aligned}
$$

Thus it suffices to show that

$$
(5.19) \qquad \int_s^r \frac{\partial}{\partial t}\eta_H(r,t)Z_H(t,s)dt = -1\,.
$$

From the definition of $\eta_H(r,t)$ and $Z_H(t,s)$, we have

$$
\int_s^r \frac{\partial}{\partial t}\eta_H(r,t)Z_H(t,s)dt = H(1-2H)\kappa_1\left[I_1 + I_2 + I_3\right]\,,
$$

where I_1, I_2, and I_3 are defined and calculated as follows.

$$
\begin{aligned}
I_1 &:= \left(\frac{1}{2}-H\right)\int_s^r t^{-\frac{1}{2}-H}\int_t^r v^{H-\frac{1}{2}}(v-t)^{-H-\frac{1}{2}}dv\; t^{H-\frac{1}{2}}s^{\frac{1}{2}-H}(t-s)^{H-\frac{1}{2}}dt \\
&= \left(\frac{1}{2}-H\right)s^{\frac{1}{2}-H}\int_s^r v^{H-\frac{1}{2}}\int_s^v t^{-1}(v-t)^{-H-\frac{1}{2}}(t-s)^{H-\frac{1}{2}}dt\,dv\,.
\end{aligned}
$$

Lemma 14.4 with $\mu = \frac{1}{2}+H$, $\nu = \frac{1}{2}-H$ yields

$$
\begin{aligned}
I_1 &= \left(\frac{1}{2}-H\right)s^{\frac{1}{2}-H}B(\frac{1}{2}+H,\frac{1}{2}-H)\int_s^r s^{H-\frac{1}{2}}v^{-1}dv \\
&= \left(\frac{1}{2}-H\right)B(\frac{1}{2}+H,\frac{1}{2}-H)\log\left(\frac{r}{s}\right)\,.
\end{aligned}
$$

I_2 is defined and calculated as

$$
\begin{aligned}
I_2 &:= -\int_s^r r^{H+\frac{1}{2}}t^{-H-\frac{1}{2}}(r-t)^{-H-\frac{1}{2}}t^{H-\frac{1}{2}}s^{\frac{1}{2}-H}(t-s)^{H-\frac{1}{2}}dt \\
&= -r^{H+\frac{1}{2}}s^{\frac{1}{2}-H}\int_s^r t^{-1}(r-t)^{-H-\frac{1}{2}}(t-s)^{H-\frac{1}{2}}dt\,.
\end{aligned}
$$

Applying Lemma 14.4 with $\mu = \frac{1}{2}+H$, $\nu = \frac{1}{2}-H$, we obtain

$$
I_2 = -B(H+\frac{1}{2},\frac{1}{2}-H)\,.
$$

Finally we have

$$
\begin{aligned}
I_3 &:= -\frac{\left(H-\frac{1}{2}\right)\kappa_H}{H(1-2H)\kappa_1}\int_s^r \frac{\partial}{\partial t}\eta_H(r,t)s^{\frac{1}{2}-H}\int_s^t u^{H-\frac{3}{2}}(u-s)^{H-\frac{1}{2}}du\,dt \\
&= \frac{\left(H-\frac{1}{2}\right)\kappa_H}{H(1-2H)\kappa_1}\int_s^r \eta_H(r,t)s^{\frac{1}{2}-H}t^{H-\frac{3}{2}}(t-s)^{H-\frac{1}{2}}dt \\
&= \left(H-\frac{1}{2}\right)\int_s^r t^{\frac{1}{2}-H}\int_t^r v^{H-\frac{1}{2}}(v-t)^{-H-\frac{1}{2}}dv\,s^{\frac{1}{2}-H}t^{H-\frac{3}{2}}(t-s)^{H-\frac{1}{2}}dt \\
&= \left(H-\frac{1}{2}\right)s^{\frac{1}{2}-H}\int_s^r v^{H-\frac{1}{2}}\int_s^v t^{-1}(v-t)^{-H-\frac{1}{2}}(t-s)^{H-\frac{1}{2}}dt\,.
\end{aligned}
$$

Applying Lemma 14.4 with $\mu = \frac{1}{2} + H$, $\nu = \frac{1}{2} - H$, we obtain

$$
\begin{aligned}
I_3 &= \left(H - \frac{1}{2}\right) B(\frac{1}{2} - H, \frac{1}{2} + H) s^{\frac{1}{2}-H} \int_s^r v^{H-\frac{1}{2}} s^{H-\frac{1}{2}} v^{-H-\frac{1}{2}} dv \\
&= \left(H - \frac{1}{2}\right) B(\frac{1}{2} - H, \frac{1}{2} + H) \log\left(\frac{r}{s}\right) = -I_1 \,.
\end{aligned}
$$

Therefore

$$
\int_s^r \frac{\partial}{\partial t} \eta_H(r,t) Z_H(t,s) dt = H(1-2H)\kappa_1 I_2
$$

which is easily seen to be -1 from the definition of κ_1. This proves Theorem 5.6.
$\square$

5.6. $\Gamma_{H,T}\mathbb{B}_{H,T}$ for $0 < H < 1/2$

Now we study $\Gamma_H \mathbb{B}_{H,T}$ for $0 < H < 1/2$.

THEOREM 5.7. *Let $0 < H < \frac{1}{2}$ and let f be differentiable with $f(0) = 0$. Then*

$$
g(t) = \mathbb{B}_{H,T} f(t) = \frac{d}{dt} \int_0^t \eta_H(t,s) f(s) ds \,, \quad 0 \le t \le T
$$

is well-defined and $\Gamma_H g$ is also well-defined. Moreover,

$$
(5.20) \qquad \Gamma_H g(r) = \frac{d}{dr} \int_0^r Z_H(r,t) g(t) dt = f(r) \,, \quad 0 \le t \le T \,.
$$

Proof First we have

$$
\begin{aligned}
\Gamma_H g(r) &= \frac{d}{dr} \int_0^r Z_H(r,t) g(t) dt \\
&= \frac{d}{dr} \int_0^r Z_H(r,t) \frac{d}{dt} \int_0^t \eta_H(t,s) f(s) ds \, dt \\
&= \frac{d}{dr} \int_0^r Z_H(r,t) \int_0^t \frac{\partial}{\partial t} \eta_H(t,s) f(s) ds \, dt \,.
\end{aligned}
$$

Thus it suffices to show that

$$
\int_s^r Z_H(r,t) \frac{\partial}{\partial t} \eta_H(t,s) dt = 1 \,.
$$

By the definition of $\eta_H(t,s)$ we have

$$
\frac{\partial}{\partial t} \eta_H(t,s) = \frac{H(1-2H)\kappa_1}{\kappa_H} s^{\frac{1}{2}-H} t^{H-\frac{1}{2}} (t-s)^{-\frac{1}{2}-H} \,.
$$

From the definition of $Z_H(r,t)$ it follows that

$$
\int_s^r Z_H(r,t) \frac{\partial}{\partial t} \eta_H(t,s) dt = H(1-2H)\kappa_1(I_1 + I_2) \,,
$$

where

$$
\begin{aligned}
I_1 &:= \int_s^r r^{H-\frac{1}{2}} t^{\frac{1}{2}-H} (r-t)^{H-\frac{1}{2}} s^{\frac{1}{2}-H} t^{H-\frac{1}{2}} (t-s)^{-\frac{1}{2}-H} dt \\
&= r^{H-\frac{1}{2}} s^{\frac{1}{2}-H} \int_s^r (t-s)^{-\frac{1}{2}-H} (r-t)^{H-\frac{1}{2}} dt \\
&= r^{H-\frac{1}{2}} s^{\frac{1}{2}-H} B(\frac{1}{2} + H, \frac{1}{2} - H) \,.
\end{aligned}
$$

And

$$
\begin{aligned}
I_2 &:= -\left(H - \frac{1}{2}\right) \int_s^r t^{\frac{1}{2}-H} \int_t^r u^{H-\frac{3}{2}}(u-t)^{H-\frac{1}{2}} du\, s^{\frac{1}{2}-H} t^{H-\frac{1}{2}}(t-s)^{-\frac{1}{2}-H} dt \\
&= -\left(H - \frac{1}{2}\right) s^{\frac{1}{2}-H} \int_s^r u^{H-\frac{3}{2}} \int_s^u (t-s)^{-\frac{1}{2}-H}(u-t)^{H-\frac{1}{2}} dt\, du \\
&= -\left(H - \frac{1}{2}\right) B(\frac{1}{2}+H, \frac{1}{2}-H) s^{\frac{1}{2}-H} \int_s^r u^{H-\frac{3}{2}} du \\
&= B(\frac{1}{2}+H, \frac{1}{2}-H)\left[1 - r^{H-\frac{1}{2}} s^{\frac{1}{2}-H}\right].
\end{aligned}
$$

Therefore

$$
\int_s^r Z_H(r,t) \frac{\partial}{\partial t} \eta_H(t,s) dt = H(1-2H)\kappa_1 B(\frac{1}{2}+H, \frac{1}{2}-H) = 1.
$$

This proves the theorem. □

5.7. Transpose of $\mathbb{I}_{H,T}$

The following theorem shows that if $1/2 < H < 1$, then the transpose of $\mathbb{I}_{H,T}$ restricted to the interval $[0,T]$ is

$$(5.21)\quad \mathbb{I}_{H,T}^* f(t) := (H - \frac{1}{2})\kappa_H t^{\frac{1}{2}-H} \int_t^T u^{H-\frac{1}{2}}(u-t)^{H-\frac{3}{2}} f(u) du, \quad 0 \le t \le T.$$

When there is no ambiguity, we denote $\mathbb{I}_H^* = \mathbb{I}_{H,T}^*$. Denote by $\mathbf{S} = \mathbf{S}[0,T]$ the set of all smooth function on $[0,T]$ with bounded derivatives of all orders.

THEOREM 5.8. Let $1/2 < H < 1$ and let $\mathbb{I}_H^*$ be defined by (5.21). For any function $f, g \in \mathbf{S}$, we have

$$(5.22)\qquad \int_0^T [\mathbb{I}_H f(t)] g(t) dt = \int_0^T f(t) [\mathbb{I}_H^* g(t)]\, dt.$$

Proof It is easy to check the following

$$
\begin{aligned}
\int_0^T [\mathbb{I}_H f(t)] g(t) dt &= (H-\frac{1}{2})\kappa_H \int_0^T \int_0^t s^{\frac{1}{2}-H} t^{H-\frac{1}{2}}(t-s)^{H-\frac{3}{2}} f(s) ds\, g(t) dt \\
&= (H-\frac{1}{2})\kappa_H \int_0^T s^{\frac{1}{2}-H} \int_s^T t^{H-\frac{1}{2}}(t-s)^{H-\frac{3}{2}} g(t) dt\, f(s) ds.
\end{aligned}
$$

This shows (5.22). □

In general, the transpose of this operator $\mathbb{I}_H$ is easily found to be

$$(5.23)\qquad \mathbb{I}_H^* g(s) = Z_H(T,s)g(T) - \int_s^T Z_H(t,s)g'(t) dt.$$

This expression can be simplified.

LEMMA 5.9. Let $0 < H < 1$. If $g \in C_b^\infty([0,T])$, then

$$(5.24)\qquad \mathbb{I}_H^* g(s) = -\kappa_H s^{\frac{1}{2}-H} \frac{d}{ds} \int_s^T t^{H-\frac{1}{2}}(t-s)^{H-\frac{1}{2}} g(t) dt, \quad 0 \le s \le T.$$

Proof   Make substitution $t = u + s$.

$$\frac{d}{ds} \int_s^T t^{H-\frac{1}{2}} (t-s)^{H-\frac{1}{2}} g(t) dt$$

$$= \frac{d}{ds} \int_0^{T-s} (u+s)^{H-\frac{1}{2}} u^{H-\frac{1}{2}} g(u+s) du$$

$$= -T^{H-\frac{1}{2}} (T-s)^{H-\frac{1}{2}} g(T) + \int_0^{T-s} (u+s)^{H-\frac{1}{2}} u^{H-\frac{1}{2}} g'(u+s) du$$

$$+ (H - \frac{1}{2}) \int_0^{T-s} (u+s)^{H-\frac{3}{2}} u^{H-\frac{1}{2}} g(u+s) du$$

$$= -T^{H-\frac{1}{2}} (T-s)^{H-\frac{1}{2}} g(T) + \int_s^T t^{H-\frac{1}{2}} (t-s)^{H-\frac{1}{2}} g'(t) dt$$

$$(5.25) \qquad + (H - \frac{1}{2}) \int_s^T t^{H-\frac{3}{2}} (t-s)^{H-\frac{1}{2}} g(t) dt \,.$$

On the other hand, it is easy to check that

$$s^{\frac{1}{2}-H} \int_s^T \left[\int_s^t u^{H-\frac{3}{2}} (u-s)^{H-\frac{1}{2}} du \right] g'(t) dt = g(T) s^{\frac{1}{2}-H} \int_s^T u^{H-\frac{3}{2}} (u-s)^{H-\frac{1}{2}} du$$

$$- s^{\frac{1}{2}-H} \int_s^T t^{H-\frac{3}{2}} (t-s)^{H-\frac{1}{2}} g(t) dt \,.$$

Namely,

$$s^{\frac{1}{2}-H} \int_s^T t^{H-\frac{3}{2}} (t-s)^{H-\frac{1}{2}} g(t) dt = g(T) s^{\frac{1}{2}-H} \int_s^T u^{H-\frac{3}{2}} (u-s)^{H-\frac{1}{2}} du$$

$$(5.26) \qquad - s^{\frac{1}{2}-H} \int_s^T \left[\int_s^t u^{H-\frac{3}{2}} (u-s)^{H-\frac{1}{2}} du \right] g'(t) dt \,.$$

Replacing the last term of (5.25) by (5.26) and noticing the expression for $Z_H(t,s)$ given by (5.3), we obtain (5.24) from (5.23). The lemma is proved. $\qquad \square$

The following form of $\mathbb{I}_{H,T}^*$ is also interesting, which is known for general kernel (see Lemma 1 of [**3**]).

LEMMA 5.10. *If g is Hölder continuous of exponent α with $\alpha > 1/2 - H$ (if $1/2 < H < 1$, then we assume that g is continuous), then*

$$(5.27) \qquad \mathbb{I}_H^* g(s) = Z_H(T,s) g(s) + \int_s^T [g(t) - g(s)] \frac{\partial Z_H}{\partial t}(t,s) dt \,.$$

Proof   Applying integration by parts formula to $[g(t) - g(s)] Z_H(t,s)$ in the integration of (5.23), and noticing $\lim_{t \downarrow s} [g(t) - g(s)] Z_H(t,s) = 0$, we obtain (5.27). $\square$

EXAMPLE 5.11. *Let $H \neq 1/2$ and $0 \le a < b \le T$. Then*

$$\mathbb{I}_{H,T}^* \chi_{[a,b]}(s) = Z_H(b,s) - Z_H(a,s) \,.$$

Proof By linearity we may assume that $a = 0$. It follows from (5.24) that

$$
\begin{aligned}
\mathbb{I}^*_{H,T}\chi_{[0,b]}(s) &= -\kappa_H s^{\frac{1}{2}-H}\frac{d}{ds}\int_s^b t^{H-\frac{1}{2}}(t-s)^{H-\frac{1}{2}}dt \\
&= -\kappa_H s^{\frac{1}{2}-H}\frac{d}{ds}\int_0^{b-s}(\xi+s)^{H-\frac{1}{2}}\xi^{H-\frac{1}{2}}dt .
\end{aligned}
$$

Now it is easy to check that $\mathbb{I}^*_{H,T}\chi_{[0,b]}(s) = Z_H(b,s)$. □

5.8. The Expression for $\mathbb{I}_{H,T}\mathbb{I}^*_{H,T}$

Let us compute $\mathbb{I}_H\mathbb{I}^*_H$.

THEOREM 5.12. *Let $1/2 < H < 1$. For any function $f \in \mathbf{S}$, we have*

$$
(5.28)\qquad \mathbb{I}_H\mathbb{I}^*_H f(t) = H(2H-1)\int_0^T |t-s|^{2H-2}f(s)ds, \quad 0 \le t \le T .
$$

Proof Let $f \in \mathbf{S}$. From the expression (5.21) and (5.4)-(5.5) it follows that

$$
\begin{aligned}
\mathbb{I}_H\mathbb{I}^*_H f(t) &= (H-\tfrac{1}{2})^2\kappa_H^2 \int_0^t t^{H-\frac{1}{2}}s^{\frac{1}{2}-H}(t-s)^{H-\frac{3}{2}}s^{\frac{1}{2}-H} \\
&\qquad\qquad \int_s^T u^{H-\frac{1}{2}}(u-s)^{H-\frac{3}{2}}f(u)du\,ds \\
&= (H-\tfrac{1}{2})^2\kappa_H^2 t^{H-\frac{1}{2}}\int_0^T\int_0^{u\wedge t} s^{1-2H}(t-s)^{H-\frac{3}{2}} \\
&\qquad\qquad (u-s)^{H-\frac{3}{2}}ds\, u^{H-\frac{1}{2}}f(u)du .
\end{aligned}
$$

Applying Lemma 14.3 we have

$$
\begin{aligned}
\mathbb{I}_H\mathbb{I}^*_H f(t) &= (H-\tfrac{1}{2})^2\kappa_H^2 t^{H-\frac{1}{2}}B(2-2H, H-\tfrac{1}{2}) \\
&\qquad\qquad \int_0^T u^{\frac{1}{2}-H}t^{\frac{1}{2}-H}|t-u|^{2H-2}u^{H-\frac{1}{2}}f(u)du \\
(5.29)\qquad &= H(2H-1)\int_0^T |t-u|^{2H-2}f(u)du ,
\end{aligned}
$$

proving the theorem. □

If $0 < H < 1/2$ and $g \in \mathbf{S}$, then

$$
\begin{aligned}
\mathbb{I}_H \mathbb{I}_H^* g(t) &= -\frac{d}{dt} \int_0^t Z_H(t,s) \int_s^T Z_H(u,s) g'(u) du \, ds \\
&\quad + \frac{d}{dt} \int_0^t Z_H(t,s) Z_H(T,s) g(T) ds \\
&= -\frac{1}{2} \frac{d}{dt} \int_0^T \int_0^{u \wedge t} Z_H(t,s) Z_H(u,s) ds \, g'(u) du \\
&\quad + \frac{1}{2} \frac{d}{dt} \left[T^{2H} + t^{2H} + |T - t|^{2H} \right] g(T) \\
&= -\frac{1}{2} \frac{d}{dt} \int_0^T \left[u^{2H} + t^{2H} - |u - t|^{2H} \right] g'(u) du \\
&\quad + \frac{1}{2} \frac{d}{dt} \left[T^{2H} + t^{2H} - |T - t|^{2H} \right] g(T) \\
&= H t^{2H-1} g(0) + H |T - t|^{2H-1} g(T) \\
&\quad + H \int_0^T |t - u|^{2H-1} \mathrm{sign}\,(t - u) g'(u) du \,.
\end{aligned}
$$

(5.30)

Of course (5.30) is also true for $1/2 < H < 1$ from an integration by parts.

If $0 < H < 1/2$, we may also use integration by parts formula to obtain

$$
\begin{aligned}
\mathbb{I}_H \mathbb{I}_H^* g(t) &= H t^{2H-1} g(0) + H |T - t|^{2H-1} g(T) - \frac{1}{2} t^{2H} g'(0) \\
&\quad + \frac{1}{2} (T - t)^{2H} g'(T) - \frac{1}{2} \int_0^T |t - u|^{2H} g''(u) du \,.
\end{aligned}
$$

(5.31)

5.9. The transpose of $\mathbb{B}_{H,T}$

If $0 < H < 1/2$, then

$$
\begin{aligned}
\int_0^T g(t) \mathbb{B}_{H,T} f(t) dt &= \int_0^T g(t) \frac{d}{dt} \int_0^t \eta_H(t,s) f(s) ds \\
&= \int_0^T g(t) \frac{H(1-2H)\kappa_1}{\kappa_H} \int_0^t s^{\frac{1}{2}-H} t^{H-\frac{1}{2}} (t-s)^{-\frac{1}{2}-H} f(s) ds \\
&= \frac{H(1-2H)\kappa_1}{\kappa_H} \int_0^T \int_s^T t^{H-\frac{1}{2}} (t-s)^{-\frac{1}{2}-H} g(t) dt \, s^{\frac{1}{2}-H} f(s) ds \,.
\end{aligned}
$$

This implies that

(5.32) $$ \mathbb{B}_{H,T}^* g(s) = \frac{H(1-2H)\kappa_1}{\kappa_H} s^{\frac{1}{2}-H} \int_s^T t^{H-\frac{1}{2}} (t-s)^{-\frac{1}{2}-H} g(t) dt \,. $$

If $1/2 < H < 1$, then

$$
\begin{aligned}
\int_0^T g(t) \mathbb{B}_{H,T} f(t) dt &= \int_0^T g(t) \frac{d}{dt} \int_0^t \eta_H(t,s) f(s) ds \\
&= g(T) \int_0^T \eta_H(T,t) f(t) dt - \int_0^T g'(r) \int_0^r \eta_H(r,t) f(t) dt \, dr \\
&= g(T) \int_0^T \eta_H(T,t) f(t) dt - \int_0^T \int_t^T \eta_H(r,t) g'(r) dr \, f(t) dt \,.
\end{aligned}
$$

Therefore

$$(5.33) \qquad \mathbb{B}_{H,T}^* g(t) = g(T)\eta_H(T,t) - \int_t^T \eta_H(r,t)g'(r)dr.$$

Summarizing the above argument we obtain

THEOREM 5.13. *If $0 < H < 1/2$, then the transpose of $\mathbb{B}_{H,T}$ is given by (5.32). If $1/2 < H < 1$, then the transpose of $\mathbb{B}_{H,T}$ is given by (5.33).*

If $1/2 < H < 1$, the following alternative compact form for $\mathbb{B}_{H,T}$ will be useful.

THEOREM 5.14. *Let $1/2 < H < 1$. If f is continuously differentiable, then*

$$(5.34) \qquad \mathbb{B}_{H,T} f(t) = \frac{2H\kappa_1}{\kappa_H} t^{H-\frac{1}{2}} \frac{d}{dt} \int_0^t r^{\frac{1}{2}-H}(t-r)^{\frac{1}{2}-H} f(r)dr.$$

Proof By definition

$$
\begin{aligned}
\mathbb{B}_{H,T} f(t) &= \frac{d}{dt} \int_0^t \eta_H(t,r) f(r)dr \\
&= \frac{2H\kappa_1}{\kappa_H} \frac{d}{dt}\left[t^{H-\frac{1}{2}} \int_0^t r^{\frac{1}{2}-H}(t-r)^{\frac{1}{2}-H} f(r)dr \right] \\
&\quad -(H-\frac{1}{2})\frac{2H\kappa_1}{\kappa_H} \frac{d}{dt}\left[\int_0^t \int_r^t v^{H-\frac{3}{2}} r^{\frac{1}{2}-H}(v-r)^{\frac{1}{2}-H} dv f(r)dr \right] \\
&= \frac{2H\kappa_1}{\kappa_H}\left[\frac{d}{dt} t^{H-\frac{1}{2}} \int_0^t r^{\frac{1}{2}-H}(t-r)^{\frac{1}{2}-H} f(r)dr \right. \\
&\quad \left. -(H-\frac{1}{2})\int_0^t t^{H-\frac{3}{2}} r^{\frac{1}{2}-H}(t-r)^{\frac{1}{2}-H} f(r)dr \right] \\
&= \frac{2H\kappa_1}{\kappa_H} t^{H-\frac{1}{2}} \frac{d}{dt} \int_0^t r^{\frac{1}{2}-H}(t-r)^{\frac{1}{2}-H} f(r)dr.
\end{aligned}
$$

This proves the theorem.   □

Corresponding to this form we have an alternative expression for $\mathbb{B}_{H,T}^*$.

THEOREM 5.15. *If $1/2 < H < 1$ and if f is continuously differentiable, then*

$$(5.35) \qquad \mathbb{B}_{H,T}^* f(t) = -\frac{2H\kappa_1}{\kappa_H} t^{\frac{1}{2}-H} \frac{d}{dt} \int_t^T (u-t)^{\frac{1}{2}-H} u^{H-\frac{1}{2}} f(u)du.$$

Proof Let g be any continuously differentiable function and consider

$$
\begin{aligned}
\int_0^T g(t)\mathbb{B}^*_{H,T}f(t)dt &= \frac{2H\kappa_1}{\kappa_H}\int_0^T t^{H-\frac{1}{2}}g(t)\frac{d}{dt}\int_0^t r^{\frac{1}{2}-H}(t-r)^{\frac{1}{2}-H}f(r)drdt\\
&= \frac{2H\kappa_1}{\kappa_H}\left[T^{H-\frac{1}{2}}g(T)\int_0^T r^{\frac{1}{2}-H}(T-r)^{\frac{1}{2}-H}f(r)dr\right.\\
&\quad \left. -\int_0^T\int_0^t r^{\frac{1}{2}-H}(t-r)^{\frac{1}{2}-H}f(r)dr\frac{d}{dt}\left(t^{H-\frac{1}{2}}g(t)\right)dt\right]\\
&= \frac{2H\kappa_1}{\kappa_H}\left[T^{H-\frac{1}{2}}g(T)\int_0^T r^{\frac{1}{2}-H}(T-r)^{\frac{1}{2}-H}f(r)dr\right.\\
&\quad \left. -\int_0^T\int_r^T r^{\frac{1}{2}-H}(t-r)^{\frac{1}{2}-H}\frac{d}{dt}\left(t^{H-\frac{1}{2}}g(t)\right)dt\right]f(r)dr\,.
\end{aligned}
$$

Therefore

$$
\begin{aligned}
\mathbb{B}^*_{H,T}f(r) &= \frac{2H\kappa_1}{\kappa_H}\left[T^{H-\frac{1}{2}}r^{\frac{1}{2}-H}(T-r)^{\frac{1}{2}-H}g(T)\right.\\
&\quad \left. -r^{\frac{1}{2}-H}\int_r^T(t-r)^{\frac{1}{2}-H}\frac{d}{dt}\left(t^{H-\frac{1}{2}}g(t)\right)dt\right].
\end{aligned}
$$

It is easy to see that this is exactly (5.35). $\qquad\square$

5.10. The Expression of $\mathbb{B}^*_{H,T}\mathbb{B}_{H,T}$

To obtain the explicit form of the Cameron-Martin-Girsanov type formula for nonlinear transformation for the fBm, we need to find the inverse transformation of $\mathbb{I}_{H,T}\mathbb{I}^*_{H,T}$. We also need to compute the inverse of $\mathbb{I}_{H,T}\mathbb{I}^*_{H,T}$ in many other applications. It is easy to check that $(\mathbb{I}_{H,T}\mathbb{I}^*_{H,T})^{-1} = (\mathbb{I}^*_{H,T})^{-1}(\mathbb{I}_{H,T})^{-1} = \mathbb{B}^*_{H,T}\mathbb{B}_{H,T}$. Now we shall obtain the explicit form for $\mathbb{B}^*_{H,T}\mathbb{B}_{H,T}$.

THEOREM 5.16. *If $1/2 < H < 1$, then*

$$
\begin{aligned}
\mathbb{B}^*_{H,T}\mathbb{B}_{H,T}f(t) &= -\zeta_H t^{1/2-H}\frac{d}{dt}\int_t^T dw\, w^{2H-1}(w-t)^{1/2-H}\\
&\quad \times\frac{d}{dw}\int_0^w dz\, z^{1/2-H}(w-z)^{1/2-H}f(z)\,,
\end{aligned}
$$
(5.36)

where

$$
\zeta_H = \frac{4H^2\kappa_1^2}{\kappa_H^2} = \frac{\Gamma(2-2H)}{2H\Gamma(\frac{1}{2}+H)\Gamma(\frac{3}{2}-H)^3}\,.
$$
(5.37)

If $0 < H < 1/2$, then

$$
\mathbb{B}^*_{H,T}\mathbb{B}_{H,T}f(t) = \int_0^T \chi_H(T,t,u)f(u)du\,,
$$
(5.38)

where

$$
\chi_H(T,t,u) = \tilde{\zeta}_H(tu)^{\frac{1}{2}-H}\int_{t\wedge u}^T r^{2H-1}(r-t)^{-\frac{1}{2}-H}(r-u)^{-\frac{1}{2}-H}dr
$$
(5.39)

and

$$\tilde{\zeta}_H = \frac{(1-2H)^2 \zeta_H}{4} = \frac{(1-2H)^2 \Gamma(2-2H)}{8H\Gamma(\frac{1}{2}+H)\Gamma(\frac{3}{2}-H)^3}.$$

Proof If $1/2 < H < 1$, then (5.36)-(5.37) are direct consequence of (5.34) and (5.35).

Now let $0 < H < 1/2$. From the formulas (5.8) and (5.32) of $\mathbb{B}_{H,T}$ and $\mathbb{B}^*_{H,T}$, it follows that

$$
\begin{aligned}
\mathbb{B}^*_{H,T}\mathbb{B}_{H,T}f(t) &= \tilde{\zeta}_H t^{\frac{1}{2}-H} \int_t^T r^{H-\frac{1}{2}}(r-t)^{-\frac{1}{2}-H} \int_0^r u^{\frac{1}{2}-H} r^{H-\frac{1}{2}}(r-u)^{-\frac{1}{2}-H}f(u)\,du\,dr \\
&= \tilde{\zeta}_H t^{\frac{1}{2}-H} \int_0^t u^{\frac{1}{2}-H} \int_t^T (r-t)^{-\frac{1}{2}-H} r^{2H-1}(r-u)^{-\frac{1}{2}-H}\,dr\,f(u)\,du \\
&\quad + \tilde{\zeta}_H t^{\frac{1}{2}-H} \int_t^T u^{\frac{1}{2}-H} \int_u^T (r-t)^{-\frac{1}{2}-H} r^{2H-1}(r-u)^{-\frac{1}{2}-H}\,dr\,f(u)\,du \\
&= \int_0^T \chi_H(T,t,u)f(u)\,du.
\end{aligned}
$$

This shows the theorem. $\square$

REMARK 5.17. If $1/2 < H < 1$, then we have that $g(t) = \mathbb{B}^*_{H,T}\mathbb{B}_{H,T}f(t)$ is the solution of the following integral equation

$$H(2H-1)\int_0^T (t-s)^{2H-2}g(s)\,ds = f(t), \quad 0 \le t \le T.$$

This equation is called Carleman equation and has been studied extensively (see for example, [106] and [76]). It is applied to obtain prediction theorem and Girsanov theorem in [44], [47], [77]. The method presented here through (5.34) and (5.35) is more elementary and uses only some elementary integral identity.

Let $0 < H < 1/2$ and consider the case $f = 1$. We have

$$
\begin{aligned}
\left[\mathbb{B}^*_{H,T}\mathbb{B}_{H,T}1\right](s) &= \tilde{\zeta}_H t^{\frac{1}{2}-H} \int_t^T r^{H-\frac{1}{2}}(r-t)^{-\frac{1}{2}-H} r^{H-\frac{1}{2}} \int_0^r u^{\frac{1}{2}-H}(r-u)^{-\frac{1}{2}-H}\,du\,dr \\
&= \tilde{\zeta}_H t^{\frac{1}{2}-H} \int_t^T r^{2H-1}(r-t)^{-\frac{1}{2}-H} B(\frac{3}{2}-H, \frac{1}{2}-H) r^{1-2H}\,dr \\
&= \tilde{\zeta}_H t^{\frac{1}{2}-H} B(\frac{3}{2}-H, \frac{1}{2}-H) \frac{1}{\frac{1}{2}-H}(T-t)^{\frac{1}{2}-H}.
\end{aligned}
$$

It is straightforward to check that

$$\tilde{\zeta}_H B(\frac{3}{2}-H, \frac{1}{2}-H)\frac{1}{\frac{1}{2}-H} = \frac{1}{2H\Gamma(\frac{3}{2}-H)}.$$

Thus we have

PROPOSITION 5.18. Let $0 < H < 1/2$. The following identity is true:

$$(5.40) \qquad \left[\mathbb{B}^*_{H,T}\mathbb{B}_{H,T}1\right](s) = \frac{1}{2H\Gamma(\frac{3}{2}-H)} t^{\frac{1}{2}-H}(T-t)^{\frac{1}{2}-H}.$$

5.11. Extension of $\mathbb{I}\!\Gamma^*_{H,T}$ and $\mathbb{B}^*_{H,T}$

Let $\mathbf{A} := \big\{ \mathbb{B}^*_{H,T} f : \ f \in \mathbf{S} \big\}$, where $\mathbf{S}$ denotes the set of all smooth functions on $[0,T]$ whose derivatives are bounded. If $\mathbb{B}^*_{H,T} f \equiv 0$, then $f = \mathbb{I}\!\Gamma^*_{H,T} \mathbb{B}^*_{H,T} f = 0$. This means that $\mathbb{B}^*_{H,T}$ is a bijection from $\mathbf{S}$ to $\mathbf{A}$. For any two elements in $\mathbf{A}$, say $g_1 = \mathbb{B}^*_{H,T} f_1$ and $g_2 = \mathbb{B}^*_{H,T} f_2$, define

$$(5.41) \qquad \langle g_1 , g_2 \rangle_{\Theta_H} := \int_0^T f_1(t) f_2(t) dt = \int_0^T \mathbb{I}\!\Gamma^*_{H,T} g_1(t) \mathbb{I}\!\Gamma^*_{H,T} g_2(t) dt .$$

Then $\mathbf{A}$ is a pre-Hilbert space with respect to the above scalar product. The completion of $\mathbf{A}$ with respect to the scalar product $\langle \cdot , \cdot \rangle_{\Theta_H}$ is denoted by $\Theta_H([0,T])$, which is a Hilbert space. Since in most part of this paper $T > 0$ is fixed we use Θ_H to denote $\Theta_H([0,T])$ (In [47], [48], we have used the same symbol to denote a similar space for $T = \infty$). $\mathbb{B}^*_{H,T} : \mathbf{S} \to \mathbf{A}$ can be extended to an isometry from the Hilbert spaces $L^2([0,T])$ to the Hilbert space Θ_H. Its inverse is the extension of the $\mathbb{I}\!\Gamma^*_{H,T}$ defined by (5.30) or (5.31) to Θ_H. We continue to use $\mathbb{B}^*_{H,T}$ and $\mathbb{I}\!\Gamma^*_{H,T}$ to denote their extensions.

Summarizing, we state

LEMMA 5.19. $\mathbb{B}^*_{H,T} : \ L^2([0,T]) \to \Theta_H$ is an isometry from the Hilbert space $L^2([0,T])$ to the Hilbert space Θ_H. The value of $\mathbb{B}^*_{H,T}$ on smooth function space $\mathbf{S}$ is given by (5.32), (5.33), or (5.35). The inverse of $\mathbb{B}^*_{H,T}$ is an isometry, denoted by $\mathbb{I}\!\Gamma^*_{H,T} : \Theta_H([0,T]) \to L^2([0,T])$. The value of $\mathbb{I}\!\Gamma^*_{H,T}$ on smooth function space $\mathbf{S}$ is given by (5.21), (5.23), (5.24), or (5.27).

It is known that when $H > 1/2$, Θ_H contains distributions. To avoid dealing with distributions other spaces are introduced for instance in [92]. One of them is the Banach space

$$\hat{\Theta}_H := \left\{ f ; \ \int_0^T \int_0^T |f(u)||f(v)||u - v|^{2H-2} du dv < \infty \right\} .$$

It is obvious that this space is a subspace of Θ_H. As illustrated in [47] and [48], sometimes, Θ_H is more convenient in stochastic analysis.

Since $\mathbb{B}^*_{H,T} : L^2([0,T]) \to \Theta_H$ is an isometry its transpose $(\mathbb{B}^*_{H,T})^*$ is an isometry from Θ^*_H to $(L^2([0,T]))^*$, where Θ^*_H is the dual of the Hilbert space Θ_H and $(L^2([0,T]))^*$ is the dual of the Hilbert space $L^2([0,T])$. We may identify the element of $\Theta^*_H \cap L^2([0,T])$ in the following way: For any $h \in L^2([0,T])$ consider the following linear functional on $\mathbf{S}$:

$$h(f) = \int_0^T h(t) f(t) dt , \quad f \in \mathbf{S} .$$

If this functional can be extended to a continuous linear functional on Θ_H, then h can be considered as an element of Θ^*_H. It is easy to check that if $h \in \mathbf{S}$, then h can be considered as an element of Θ^*_H. We can, in this way, embed $\mathbf{S}$ into Θ^*_H by the duality that we identify $(L^2([0,T]))^*$ (the dual of the Hilbert space $L^2([0,T])$) as $L^2([0,T])$:

$$(L^2([0,T]))^* = L^2([0,T]) .$$

The meaning can be seen more clearly from the following lemma. Let us introduce the following Hilbert space

$$W^{1,2}([0,T]) = \left\{ f \, ; \, \int_0^T |f(t)|^2 dt + \int_0^T |f'(t)|^2 dt < \infty \right\} .$$

LEMMA 5.20. *(a) If $0 < H < 1/2$, then $W^{1,2}([0,T]) \subset \Theta_H([0,T])$. Therefore $\Theta_H([0,T])^* \subset W^{-1,2}([0,T])$, where $W^{1,2}([0,T])$ is the Sobolev space over $[0,T]$ and $W^{-1,2}([0,T])$ is dual of $W^{1,2}([0,T])$.*

(b) If $1/2 < H < 1$, then $L^2([0,T]) \subset \Theta_H([0,T])$. Therefore $\Theta_H([0,T])^* \subset L^2([0,T])$.

Consequently, $\mathbf{S}$ is a dense subset of $\Theta_H([0,T])$ and that of $\Theta_H([0,T])^*$ for all H.

Proof Let us consider the case $0 < H < 1/2$. Let $f \in \mathbf{S}$ and $g = \mathbb{B}^*_{H,T} f$. Then by definition and Eq. (5.23) we have

$$
\begin{aligned}
\|g\|^2_{\Theta_H} &= \int_0^T |f(t)|^2 dt = \int_0^T |\mathbb{I}^*_{H,T} g(s)|^2 ds \\
&\leq 2 \Big(\int_0^T |Z(T,s)|^2 ds |g(T)|^2 + \int_0^T |\int_0^T \chi_{[0,t]}(s) Z_H(t,s) g'(t) dt|^2 ds \Big) .
\end{aligned}
$$

By the generalized Minkowski inequality and Sobolev inequality, we have

$$
\begin{aligned}
\|g\|^2_{\Theta_H} &\leq C\|g\|_{1,2} + C \left\{ \int_0^T \left[\int_0^T |\chi_{[0,t]}(s) Z_H(t,s) g'(t)|^2 ds \right]^{1/2} dt \right\}^2 \\
&\leq C\|g\|_{1,2} .
\end{aligned}
$$

Now consider the case $1/2 < H < 1$. We have

$$
\begin{aligned}
\|g\|^2_{\Theta_H} &= \int_0^T |f(t)|^2 dt = \int_0^T |\mathbb{I}_{H,T} g(s)|^2 ds \\
&= H(2H-1) \int_0^T \int_0^T |u-v|^{2H-2} g(u) g(v) du dv \\
&\leq C\|g\|^2_{\frac{1}{H}} ,
\end{aligned}
$$

where the last inequality follows from (2.1) of [**82**]. Therefore

$$\|g\|^2_{\Theta_H} \leq C\|g\|^2_2 .$$

The last part of the lemma is a direct consequence of (a) and (b). $\square$

5.12. Representation of Brownian motion by fractional Brownian motion

It is easy to obtain the following theorem.

THEOREM 5.21. *Let $0 < H < 1$ and let $\eta_H(t,s)$ be given by (5.7).*

(5.42) $$X_t = \int_0^t \eta_H(t,s) dB_s^H , \quad t \geq 0$$

is a standard Brownian motion.

CHAPTER 6

Stochastic Calculus of Variation

6.1. Stochastic Integral for Deterministic Integrands

Recall that if $(B_t, t \geq 0)$ is an sBm on some probability space $(\Omega, \mathcal{F}, P)$, then $B_t^H = \int_0^t Z_H(t,s) dB_s, t \geq 0$, is an fBm of Hurst parameter H on the same probability space. From the definition of $\mathbb{I}_{H,T}$ we have formally that $\dot{B}_t^H = \mathbb{I}_{H,T}(\dot{B})(t), 0 \leq t \leq T$. Heuristically, we may write

$$\int_0^T f(t) dB_t^H = \int_0^T f(t) \dot{B}_t^H dt = \int_0^T f(t) \mathbb{I}_{H,T}(\dot{B})(t) dt$$

$$(6.1) \qquad\qquad = \int_0^T (\mathbb{I}_{H,T}^* f)(t) \dot{B}_t dt = \int_0^T (\mathbb{I}_{H,T}^* f)(t) dB_t.$$

We shall use the above identity to define our stochastic integral for deterministic integrands.

Let $f \in \Theta_H$. Then it is known from the construction of Θ_H that $g = \mathbb{I}_{H,T}^* f \in L^2([0,T])$. Thus $\int_0^T g(t) dB_t$ is well-defined.

DEFINITION 6.1. Let $f \in \Theta_H$. Define

$$(6.2) \qquad\qquad \int_0^T f(t) dB_t^H := \int_0^T (\mathbb{I}_{H,T}^* f)(t) dB_t.$$

REMARK 6.2. When $1/2 < H < 1$ an element $f \in \Theta_H$ may not be a classical function of t. But it is not ambiguous to write $(f(t), 0 \leq t \leq T)$, $\int_0^T f(t) dB_t^H$, and so on. As a consequence we have

$$\int_0^T g(t) dB_t = \int_0^T \left(\mathbb{B}_{H,T}^* g \right)(t) dB_t^H.$$

It is easy to check that

PROPOSITION 6.3. Let $f \in \Theta_H$. Then $\int_0^T f(t) dB_t^H$ is a Gaussian random variable. The following identities hold

$$(6.3) \qquad\qquad \mathbb{E} \left\{ \int_0^T f(t) dB_t^H \right\} = 0$$

and

$$(6.4) \qquad\qquad \mathbb{E} \left\{ \int_0^T f_1(t) dB_t^H \int_0^T f_2(t) dB_t^H \right\} = \langle f_1, f_2 \rangle_{\Theta_H}.$$

EXAMPLE 6.4. Let $0 < a < b < T$ and let $f(t) = I_{[a,b]}(t)$ be the indicate function over interval $[a,b]$. We have that $f \in \Theta_H$ and by Example 5.11

$$(6.5) \qquad\qquad \mathbb{I}_{H,T}^* f(s) = Z_H(b,s) - Z_H(a,s).$$

Here and in what follows we understand the $Z_H(b,s) = 0$ if $s \geq b$. Therefore

$$
\begin{aligned}
\int_0^T \chi_{[a,b]}(s) dB_s^H &= \int_0^T \mathbb{I}_{H,T}^* \chi_{[a,b]}(s) dB_s \\
&= \int_0^T [Z_H(b,s) - Z_H(a,s)]\, dB_s \\
&= B_b^H - B_a^H\, .
\end{aligned}
$$

6.2. A Probability Structure Preserving Mapping

To define stochastic integral for general integrand one may also use the formula (6.2). If we do this, then the integrand on the right hand side of (6.2) should be a functional of sBm. In some papers stochastic integral is defined for integrands that are considered as a functional of sBm. However, this kind of definition of stochastic integral needs to be improved. The main problem is that the probability law of fBm and that of sBm are mutually singular and a functional of sBm may not be well-defined as a functional of fBm and vice versa. This means that a functional of fractional Brownian motion (a random variable on the probability space of fractional Brownian motion) may not be well-defined as a functional of standard Brownian motion. Moreover, even if the right hand side of (6.2), *i.e.* $\int_0^T (\mathbb{I}_{H,T}^* f)(t) dB_t$ is well-defined, it is not straightforward to consider it as a functional of fractional Brownian motion.

Thus there is a problem of how to associate a (nonlinear) functional of fBm with a given (nonlinear) functional of sBm and vice versa. A similar problem has been extensively studied for functionals of classical Brownian motions of different variances (see [**29**], [**52**]-[**54**], [**50**], [**65**], [**66**], [**102**], [**103**], and the references therein). In fact, Definition 6.1 (Equation (6.2)) yields a transformation which associates a given functional of fBm (or equivalently a random variable on the fractional Wiener space of Hurst parameter H) with a functional of sBm (or equivalently a random variable on the classical Wiener space). The nature of this correspondence is that they are the *same* functionals. If both functionals are well-defined as continuous functionals on the space $C([0,T], \mathbb{R})$ with respect to its sup norm, then these two functionals are exactly the same. However, this correspondence is only for linear functionals, *i.e.* the elements in the first chaos. We mainly use the integral kernels $Z_H(t,s)$ and $\eta_H(t,s)$, *i.e.* the fact that the fBm can be represented by sBm and the sBm can also be represented by fBm by Equations (2.3), (2.16), (2.19) and (5.42).

In this section we shall extend this correspondence to between nonlinear functionals of fBm and nonlinear functionals of sBm. We shall mainly discuss the case $0 < T < \infty$. The case $T = \infty$ can be treated in a similar way. Let us mention that a Girsanov type theorem for $T = \infty$ has already been obtained in [**48**].

We shall use the Hermite polynomials

$$
H_n(x) = (-1)^n e^{\frac{x^2}{2}} \frac{d^n}{dx^n}\left(e^{-\frac{x^2}{2}}\right), \quad n = 0, 1, 2, \cdots.
$$

The first a few Hermite polynomials are

$$
\begin{aligned}
H_0(x) &= 1, \quad H_1(x) = x, \quad H_2(x) = x^2 - 1, \\
H_3(x) &= x^3 - 3x, \quad H_4(x) = x^4 - 6x^2 + 3, \cdots
\end{aligned}
$$
(6.6)

The generating function is

$$e^{tx-\frac{t^2}{2}} = \sum_{n=0}^{\infty} \frac{t^n}{n!} H_n(x), \quad \forall\, t,\, x \in \mathbb{R}.$$

Let $\{\xi_1, \cdots, \xi_k, \cdots\}$ be a complete orthonormal system of $L^2([0,T])$ and let $\mathbb{I}_{H,T}^* \xi_k$ be well-defined. Denote

$$\eta_k(s) = \mathbb{I}_{H,T}^* \xi_k(s) = \kappa_H s^{\frac{1}{2}-H} \frac{d}{ds} \int_s^T t^{H-\frac{1}{2}}(t-s)^{H-\frac{1}{2}} \xi_k(t) dt, \quad 0 \le s \le T.$$

Then $\{\eta_1, \cdots, \eta_k, \cdots\}$ is a complete orthonormal system of $\Theta_H = \Theta_H([0,T])$. Without loss of generality, we may assume that η_k are in $\mathbf{S}$ for all $k = 1, 2, \cdots$. The *symmetric tensor product space* $\Theta_H^{\otimes n}$ of Θ_H can be defined in a standard way (see for example [45], [83]). If $f \in \Theta_H$ with $\|f\|_{\Theta_H} = 1$, then $f^{\otimes n}$ is in $\Theta_H^{\otimes n}$. We define (see [50] and the references therein) the multiple integral $I_n^{H,T}(f^{\otimes n})$ by the following formula:

$$I_n^{H,T}(f^{\otimes n}) = H_n\left(\int_0^T f(t) dB_t^H\right).$$

If $f = \sum_{k=1}^m c_k f_k^{\otimes n}$, where c_k, $k = 1, 2, \cdots, m$ are complex numbers and $f_k, k = 1, 2, \cdots, m$ are elements of Θ_H with $\|f_k\|_{\Theta_H} = 1$, then we define

$$I_n^{H,T}(f) = \sum_{\text{finite}} c_k I_n\left(f_k^{\otimes n}\right).$$

We can then define $I_n^{H,T}(f_1 \otimes f_2 \otimes \cdots \otimes f_n)$ by polarization. Finally a limiting argument enables us to define $I_n^{H,T}(f_n)$ for all $f_n \in \Theta_H^{\otimes n}$. Moreover, as in [50] we have

(6.7) $$\mathbb{E}\left(I_n^{H,T}(f)\right)^2 = n! \|f\|_{\Theta_H^{\otimes n}}^2.$$

EXAMPLE 6.5. Let $f \in \Theta_H$ (not necessarily $\|f\|_{\Theta_H} = 1$). Then it is easy to see that $f^{\otimes n} \in \Theta_H^{\otimes n}$ and

(6.8) $$I_n^{H,T}(f^{\otimes n}) = \sum_{k \le n/2} \frac{(-1)^k n! \|f\|_{\Theta_H}^{2k}}{2^k k! (n-2k)!} \left(\int_0^T f(t) dB_t^H\right)^{n-2k}.$$

The following theorem can be shown exactly in the same way as for example, in [35], [48], [50], [58].

THEOREM 6.6. *Let G be a square integrable random variable on $(\Omega, \mathcal{F}, P^H)$, i.e. $\mathbb{E}(G^2) < \infty$. Then there are $g_n \in \Theta_H^{\otimes n}$, $n = 1, 2, \cdots$, such that*

$$G = \mathbb{E}(G) + \sum_{n=1}^{\infty} \frac{1}{n!} I_n^{H,T}(g_n).$$

Moreover,

$$\mathbb{E}(G^2) = (\mathbb{E}(G))^2 + \sum_{n=1}^{\infty} \frac{1}{n!} \|g_n\|_{\Theta_H^{\otimes n}}^2.$$

Let $F \in L^2(\Omega, \mathcal{F}, P)$. By Wiener-Itô chaos expansion theorem on canonical Wiener space, we have

$$(6.9) \qquad F = \sum_{n=0}^{\infty} \frac{1}{n!} I_n(f_n), \quad f \in L^2([0,T]^n).$$

Since $f_n \in L^2([0,T]^n) = \left(L^2([0,T])\right)^{\otimes n}$, we see that $\left(\mathbb{B}_{H,T}^*\right)^{\otimes n} f_n$ is well-defined and in fact if f_n is continuously differentiable, then by (5.32) and (5.35)

$$\left(\mathbb{B}_{H,T}^*\right)^{\otimes n} f_n(s_1, \cdots, s_n)$$

$$(6.10) \quad = \begin{cases} \left(\frac{H(1-2H)\kappa_1}{\kappa_H}\right)^n s_1^{\frac{1}{2}-H} \cdots s_n^{\frac{1}{2}-H} \int_{s_1}^T \cdots \int_{s_n}^T t_1^{H-\frac{1}{2}} (t_1 - s_1)^{-\frac{1}{2}-H} \\ \qquad \cdots t_n^{H-\frac{1}{2}} (t_n - s_n)^{-\frac{1}{2}-H} f_n(t_1, \cdots, t_n) dt_1 \cdots dt_n \\[4pt] \qquad\qquad\qquad\qquad \text{if } 0 < H < 1/2 \\[8pt] (-1)^n \left(\frac{2H\kappa_1}{\kappa_H}\right)^n t_1^{\frac{1}{2}-H} \cdots t_n^{\frac{1}{2}-H} \frac{\partial^n}{\partial t_1 \cdots \partial t_n} \int_{t_1}^T \cdots \int_{t_n}^T (u_1 - t_1)^{\frac{1}{2}-H} \\ u_1^{H-\frac{1}{2}} \cdots (u_n - t_n)^{\frac{1}{2}-H} u_n^{H-\frac{1}{2}} f(u_1, \cdots, u_n) du_1 \cdots du_n \\[4pt] \qquad\qquad\qquad\qquad \text{if } 1/2 < H < 1 \end{cases}$$

Define

$$(6.11) \qquad V_{H,T} F = \sum_{n=0}^{\infty} \frac{1}{n!} I_n^{H,T} \left(\left(\mathbb{B}_{H,T}^*\right)^{\otimes n} f_n \right).$$

Since $\mathbb{B}_{H,T}^*$ is an isometry between the two Hilbert spaces $L^2([0,T])$ and $\Theta_H([0,T])$, Theorem 2.12 of [48] yields that

THEOREM 6.7. *The transformation $V_{H,T}$ defined by (6.11) is a probability structure preserving mapping from $L^2(\Omega, \mathcal{F}, P)$ to $L^2(\Omega, \mathcal{F}, P^H)$. Namely,*

$$(6.12) \quad V_{H,T}(F + G) = V_{H,T}(F) + V_{H,T}(G), \quad \forall\, F,\, G \in L^2(\Omega, \mathcal{F}, P)$$

$$(6.13) \qquad V_{H,T}(FG) = V_{H,T}(F) V_{H,T}(G), \quad \forall\, F,\, G \in L^2(\Omega, \mathcal{F}, P)$$

$$(6.14) \qquad\qquad \mathbb{E}[F] = \mathbb{E}[V_{H,T} F], \quad \forall\, F \in L^2(\Omega, \mathcal{F}, P),$$

REMARK 6.8. (a) The F and G in the above formulas can be extended to any measurable functionals (any random variable).

(b) By (6.2) F and $V_{H,T} F$ are the "same" random variable in some sense.

Since $V_{H,T} : L^2(\Omega, \mathcal{F}, P) \to L_H^2(\Omega, \mathcal{F}, P)$ is an isomorphism, the inverse $V_{H,T}^{-1}$ exists. In fact one can explicitly identify this $V_{H,T}^{-1}$. If

$$(6.15) \qquad g_n = \sum_{\text{finite}} a_{k_1 \cdots k_n} \xi_{k_1} \otimes \cdots \otimes \xi_{k_n}$$

and $\xi_k \in \mathbf{S}$ for all $k \geq 1$, then from Equation (5.24) it follows that

$$
\left[\mathbb{I}^*_{H,T}\right]^{\otimes n} g_n(s_1, \cdots, s_n)
$$

$$
= (-\kappa_H)^n \sum_{\text{finite}} a_{k_1 \cdots k_n} s_1^{\frac{1}{2} - H} \cdots s_n^{\frac{1}{2} - H} \frac{\partial^n}{\partial s_1 \cdots \partial s_n} \int_{s_1}^{T} \cdots \int_{s_n}^{T} t_1^{H - \frac{1}{2}} (t_1 - s_1)^{H - \frac{1}{2}} \cdots
$$

$$
t_n^{H - \frac{1}{2}} (t_n - s_n)^{H - \frac{1}{2}} \xi_{k_1}(t_1) \cdots \xi_{k_n}(t_n) dt_1 \cdots dt_n \, .
$$

For random variable of the form $I_n^{H,T}(g_n)$, where g_n is as above, then it is easy to verify that

$$
V_{H,T}^{-1}(I_n(g_n)) = I_n^{H,T}\left(\left[\mathbb{I}^*_{H,T}\right]^{\otimes n} g_n\right) \, .
$$

By a limiting argument, we have

THEOREM 6.9. *Let* $g_n \in S([0,T]^n)$. *Then* $\left(\mathbb{I}^*_{H,T}\right)^{\otimes n} g_n$ *exists as an element of* $\Theta_H^{\otimes n}$ *and*

$$
\left[\mathbb{I}^*_{H,T}\right]^{\otimes n} g_n(s_1, \cdots, s_n) = (-\kappa_H)^n s_1^{\frac{1}{2} - H} \cdots s_n^{\frac{1}{2} - H} \frac{\partial^n}{\partial s_1 \cdots \partial s_n}
$$

$$
\int_{s_1}^{T} \cdots \int_{s_n}^{T} t_1^{H - \frac{1}{2}} (t_1 - s_1)^{H - \frac{1}{2}} \cdots t_n^{H - \frac{1}{2}} (t_n - s_n)^{H - \frac{1}{2}}
$$

(6.16)
$$
g_n(t_1, \cdots, t_n) dt_1 \cdots dt_n \, .
$$

Moreover,

(6.17)
$$
V_{H,T}^{-1}\left(I_n^{H,T}(g_n)\right) = I_n^{H,T}\left(\left(\mathbb{I}^*_{H,T}\right)^{\otimes n} g_n\right) \, .
$$

In particular, if $f \in L^2([0,T])$ and $g \in \Theta_H$, then

$$
V_{H,T} \int_0^T f(t) dB_t = \int_0^T \mathbb{B}^*_{H,T} f(t) dB_t^H
$$

and

$$
V_{H,T}^{-1} \int_0^T g(t) dB_t^H = \int_0^T \mathbb{I}^*_{H,T} g(t) dB_t \, .
$$

6.3. Stochastic Integral for General Integrands

Now we define stochastic integral (of Itô-Skorohod type) for more general integrand.

DEFINITION 6.10. Let $f : (\Omega, \mathcal{F}, P^H) \to \Theta_H$ be measurable and suppose there exists a stochastic process $g(t, \omega)$ (based on sBm) such that for all $h \in \Theta_H^*$,

$$
V_{H,T}^{-1}\left(\langle h, f \rangle_{\Theta_H^*, \Theta_H}\right) = \langle \mathbb{B}_{H,T} h, g \rangle_{L^2} \, , \quad P\text{- for almost all } \omega \in \Omega \, ,
$$

where $\langle \cdot, \cdot \rangle_{\Theta_H^*, \Theta_H}$ denotes the pairing between Θ_H^* and Θ_H and $\langle \cdot, \cdot \rangle_{L^2}$ the L^2 pairing, *i.e.*

$$
\langle \xi, \eta \rangle_{L^2} = \int_0^T \xi(s) \eta(s) ds \, .
$$

If g is integrable in the sense of Nualart-Pardoux [87], then we say that f is integrable and define

$$
\int_0^T f dB_t^H := V_{H,T}\left(\int_0^T g(t, \omega) dB_t\right) \, .
$$

DEFINITION 6.11. Let $f : [0, T] \times (\Omega, \mathcal{F}, P^H) \to \mathbb{R}$ be jointly measurable and for almost every $\omega \in \Omega$, $f(\cdot)$ is in Θ_H. Suppose that for any $t \in [0, T]$, $V_{H,T}^{-1}f(t) = V_{H,T}^{-1}f(t, \cdot)$ exists as a functional of $(B_t, 0 \le t \le T)$. Assume that for almost every $\omega \in \Omega$, $V_{H,T}^{-1}f(\cdot)$ is in $\Theta_H([0, T])$. Thus $\mathbb{\Gamma}_{H,T}^* V_{H,T}^{-1}f(t)$ exists in $L^2([0, T])$ a.e. as a function of t. If the (Itô-Skorohod) integral $\int_0^T \mathbb{\Gamma}_{H,T}^* V_{H,T}^{-1}f(t)dB_t$ exists in the sense of Nualart-Pardoux [87], then $\int_0^T f(t)dB_t^H$ exists in the sense of above definition and

$$(6.18) \qquad \int_0^T f(t)dB_t^H = V_{H,T}\left\{\int_0^T \mathbb{\Gamma}_{H,T}^* V_{H,T}^{-1}f(t)dB_t\right\}.$$

It is easy to see that if $f \in \Theta_H \cap L^2([0, T])$ is deterministic (*i.e.* independent of ω), then the definition 6.11 coincides with Definition 6.1.

The following corresponding proposition is proved when $[0, T]$ is replaced by $\mathbb{R}_+$ in [47], [48]. Similar argument can be applied to obtain

PROPOSITION 6.12. Let $H > 1/2$ and let $f \in \Theta_H \cap L^2([0, T])$ for almost all $\omega \in \Omega$. If $\int_0^T f(t)dB_t^H$ is well-defined in the sense of [35], then $\int_0^T f(t)dB_t^H$ is well-defined in the sense of Definition 6.11 and both integrals coincide.

6.4. Malliavin Derivatives

In this subsection, we introduce the Malliavin type derivatives for nonlinear functionals of fBm. We will obtain the Meyer's inequality for fBm and an L_p estimate of stochastic integral. We will describe how the Malliavin derivatives (with respect to fractional Brownian motion and with respect to standard Brownian motion) correspond under this probability structure preserving mapping. This correspondence will be useful for computing the Radon-Nikodym derivative etc.

Let $\xi_1, \cdots, \xi_k, \cdots$ be an ONB of $L^2([0, T])$ such that $\xi_k \in \mathbf{S}$, $k = 1, 2, \cdots$. Define $\eta_k = \mathbb{\Gamma}_{H,T}^* \xi_k$. Then $\{\eta_1, \eta_2, \cdots\}$ is an ONB of Θ_H. Let $\mathbb{P}_T$ ($\mathbb{P}_{H,T}$) be the set of all polynomials of sBm (fBm of Hurst parameter H) over interval $[0, T]$. Namely, $\mathbb{P}_T$ ($\mathbb{P}_{H,T}$) contains all elements of the form

$$F(\omega) = f\left(\int_0^T \xi_1(t)dB_t, \cdots, \int_0^T \xi_n(t)dB_t\right)$$

$$\left(\text{or} \quad G(\omega) = f\left(\int_0^T \eta_1(t)dB_t^H, \cdots, \int_0^T \eta_n(t)dB_t\right)\right),$$

where f is a polynomial of n variables. Sometimes we drop the subindex T.

The Malliavin derivative D_s of a polynomial functional

$$F = f(\int_0^T \xi_1(t)dB_t, \cdots, \int_0^T \xi_n(t)dB_t)$$

of standard Brownian motion is defined as

$$D_s F = \sum_{j=1}^n \frac{\partial f}{\partial x_j}\left(\int_0^T \xi_1(t)dB_t, \cdots, \int_0^T \xi_n(t)dB_t\right)\xi_j(s), \qquad 0 \le s \le T.$$

Now let $G = g(\int_0^T \eta_1(t)dB_t^H, \cdots, \int_0^T \eta_n(t)dB_t^H)$ be a functional of fractional Brownian motion. We define the Malliavin derivative D_s^H by

$$D_s^H G = \sum_{j=1}^n \frac{\partial f}{\partial x_j} \left(\int_0^T \eta_1(t)dB_t^H, \cdots, \int_0^T \eta_n(t)dB_t^H \right) \eta_j(s), \quad 0 \le s \le T.$$

Higher order derivatives $D_{s_1,\cdots,s_k}^k$ and $D_{s_1,\cdots,s_k}^{H,k}$ can be defined in a similar way.

Let us recall the Meyer-Watanabe space $\mathbb{D}_{k,p}$. For any $F \in \mathbb{P}$, the following notation will be used

$$\|F\|_{k,p} := \|F\|_p + \sum_{l=1}^k \left[\mathbb{E} \left(\int_{[0T]^l} |D_{t_1,\cdots,t_l}F|^2 dt_1 \cdots dt_l \right)^{p/2} \right]^{1/p}.$$

It can be seen that $\|\cdot\|_{k,p}$ is a norm $p \in (1, \infty)$ and for non-negative integer k. Let $\mathbb{D}_{k,p}$ denote the Banach space obtained by completing $\mathbb{P}$ under the norm $\|\cdot\|_{k,p}$. Similar spaces for fBm can also be introduced. For any $F \in \mathbb{P}_H$, define

$$\|F\|_{H,k,p} := \mathbb{E} \left(\|F\|_{\Theta_H}^p \right)^{1/p} + \sum_{l=1}^k \left[\mathbb{E} \left(\|D^{H,l}F\|_{\Theta_H^{\otimes l}} \right)^{p/2} \right]^{1/p}.$$

It can be seen that $\|\cdot\|_{H,k,p}$ is a norm for non-negative integer k and $p \in (1, \infty)$. Let $\mathbb{D}_{k,p}^H$ denote the Banach space obtained by completing $\mathbb{P}_H$ under the norm $\|\cdot\|_{H,k,p}$.

Let us recall the definition of stochastic integral in an alternative point of view. The tensor product space $\Theta_H \times L^2(\Omega, \mathcal{F}, P^H)$ can be defined in standard way. Since $\{\eta_1, \eta_2, \cdots\}$ is an orthonormal basis of Θ_H. Let $\{F_k, k = 1, 2, \cdots\}$ be an orthonormal basis of $L^2(\Omega, \mathcal{F}, P^H)$. We also assume that $D_s^H F$ is continuously differentiable with respect to t and

$$\sup_{0 \le t \le T} |\frac{d}{dt} D_t^H F_k| < \infty.$$

It is easy to see that

$$(6.19) \qquad g_n(t, \omega) = \sum_{k=1}^n a_k \eta_k(t) F_k,$$

is an element of $\Theta_H \times L^2(\Omega, \mathcal{F}, P^H)$, where n is a given positive integer and a_k are given real numbers. It is straightforward to see that $f(t)$ is integrable in the sense of Definition 6.11. Thus $\int_0^T g_n(t)dB_t^H$ exists. Let f be any element of $\Theta_H \times L^2(\Omega, \mathcal{F}, P^H)$. We can choose a sequence of elements $g_n, n = 1, 2, \cdots$ of the form (6.19) such that g_n converges to f in $\Theta_H \times L^2(\Omega, \mathcal{F}, P^H)$. If $\int_0^T g_n(t)dB_t^H, n = 1, 2, \cdots$, is a Cauchy sequence in $L^2(\Omega, \mathcal{F}, P^H)$ and converges to a random variable G, then we say that f is integrable and define the above limit G as the stochastic integral, i.e. $\int_0^T f(t)dB_t^H := G$.

As in [86] we denote by $\mathbb{L}^{1,2}(\Omega, \mathcal{F}, P)$ the class of processes $f \in L^2([0, T] \times \Omega)$ on the probability space $(\Omega, \mathcal{F}, P)$ such that $f(t) \in D^{1,2}$ for almost all t, and that

$$\mathbb{E} \left\{ \int_0^T |f(t)|^2 dt \right\} + \mathbb{E} \left\{ \int_0^T \int_0^T (D_s u(t))^2 ds dt \right\} < \infty.$$

This means that

$$\mathbb{L}^{1,2}(\Omega, \mathcal{F}, P) = \left\{ f : \int_0^T \mathbb{E}\,|f(t)|^2 dt + \int_0^T \int_0^T \mathbb{E}\,|D_s f(t)|^2 ds dt < \infty \right\}.$$

The following proposition describes how stochastic integral transforms under the probability structure preserving mapping $V_{H,T}$ and will be useful later.

PROPOSITION 6.13. Let $f \in \mathbb{L}^{1,2}(\Omega, \mathcal{F}, P)$. Then $\mathbb{B}^*_{H,T}(V_{H,T} f)$ is integrable in the sense of Definition (6.11) and

(6.20) $$V_{H,T}\left(\int_0^T f(t) dB_t\right) = \int_0^T [\mathbb{B}^*_{H,T}(V_{H,T} f)](t) dB_t^H.$$

Proof Since for a.a. $\omega \in \Omega$, $f(\cdot, \omega) \in L^2([0,T])$, $V_{H,T} f(\cdot, \omega) \in L^2([0,T])$ for a.a. $\omega \in \Omega$. Thus

$$g(\cdot\,; \omega) := \mathbb{B}^*_{H,T}(V_{H,T} f)(\cdot, \omega) \in \Theta_H$$

and

$$\mathbb{I}^*_{H,T} g(\cdot, \omega) = (V_{H,T} f)(\cdot, \omega).$$

Consequently since $\mathbb{I}^*_{H,T}$ and $V_{H,T}$ commute, we have

$$\mathbb{I}^*_{H,T}(V_{H,T}^{-1} g) = f.$$

From (6.18), it follows

$$\begin{aligned}
\int_0^T (\mathbb{B}^*_{H,T})(V_{H,T} f))(t) dB_t^H &= \int_0^T g(t) dB_t^H \\
&= V_{H,T}\left(\int_0^T \mathbb{I}^*_{H,T}(V_{H,T}^{-1} g)(t) dB_t\right) \\
&= V_{H,T}\left(\int_0^T f(t) dB_t\right).
\end{aligned}$$

This proves the proposition. $\square$

LEMMA 6.14. *Let* $F \in \mathbb{D}_{1,2}(\Omega, \mathcal{F}, P)$. *Then*

(6.21) $$V_{H,T} D_s F = \mathbb{I}^*_{H,T}(s) D_s^H (V_{H,T} F),$$

where and in what follows $\mathbb{I}^*_{H,T}(s) D_s^H G$ *denotes the application of* $\mathbb{I}^*_{H,T}$ *to* $D_s^H G$ *as a function of s. Namely,*

(6.22) $$\mathbb{I}^*_{H,T}(s) D_s^H G = -\kappa_H s^{\frac{1}{2}-H} \frac{d}{ds} \int_s^T t^{H-\frac{1}{2}}(t-s)^{H-\frac{1}{2}} D_t^H G\, dt.$$

Proof Let $F = \exp\left(\int_0^T h(s) dB_s\right)$, where $h \in L^2([0,T])$. Then

$$D_s F = h(s) F.$$

Thus

$$V_{H,T} D_s F = h(s) \exp\left(\int_0^T \mathbb{B}^*_{H,T} h(s) dB_s^H\right).$$

On the other hand, we have

$$V_{H,T}F = \exp\left(\int_0^T \mathbb{B}_{H,T}^* h(s)dB_s^H\right).$$

Hence

$$D_s^H V_{H,T}F = (\mathbb{B}_{H,T}^* h)(s)\exp\left(\int_0^T \mathbb{B}_{H,T}^* h(s)dB_s^H\right).$$

Consequently, we obtain $V_{H,T}D_sF = \mathbb{I}_{H,T}^*(s)D_s^H V_{H,T}F$ for exponential functions. By linearity of $V_{H,T}$ and D_s, we conclude that (6.21) is true for all linear combinations of exponential functionals. The lemma is proved by a limiting argument. $\square$

LEMMA 6.15. *Let $G \in \mathbb{D}_{1,2}^H$. Then*

(6.23) $$D_t(V_{H,T}^{-1}G) = V_{H,T}^{-1}\mathbb{I}_{H,T}^*(t)D_t^H G.$$

Proof Denote $F_t = D_t(V_{H,T}^{-1}G)$. Then by Lemma 6.14 we have

$$V_{H,T}F_t = V_{H,T}D_t(V_{H,T}^{-1}G) = \mathbb{I}_{H,T}^*(t)D_t^H G.$$

Thus

$$D_t(V_{H,T}^{-1}G) = F_t = V_{H,T}^{-1}V_{H,T}F_t = V_{H,T}^{-1}\mathbb{I}_{H,T}^*(t)D_t^H G,$$

proving the lemma. $\square$

In general we have

LEMMA 6.16. *Let $F \in \mathbb{D}_{k,p}$ and $G \in \mathbb{D}_{k,p}^H$. Then*

(6.24) $$V_{H,T}D_{s_1,\cdots,s_k}^k F = \mathbb{I}_{H,T}^*(s_1)\cdots\mathbb{I}_{H,T}^*(s_k)D_{s_1,\cdots,s_k}^{H,k}(V_{H,T}F)$$

and

(6.25) $$D_{s_1,\cdots,s_k}^k\left(V_{H,T}^{-1}G\right) = V_{H,T}^{-1}\left(\mathbb{I}_{H,T}^*(s_1)\cdots\mathbb{I}_{H,T}^*(s_k)D_{s_1,\cdots,s_k}^{H,k}G\right).$$

Now let us recall the important Meyer's inequality.

For an element $F \in \mathbb{P}_H$, F can be written as the (finite) chaos expansion

$$F = \sum_{n=0}^m F_n.$$

Similar to the Ornstein-Uhlenbeck operator L in the classical Wiener space, the Ornstein-Uhlenbeck operator L_H in the fractional Wiener space is defined as

$$L_H F = \sum_{n=0}^\infty (-n)F_n.$$

Let $s \in \mathbb{R}$. In fact, we need the operator $(I - L)^{s/2}$ defined by

$$(I - L_H)^{s/2}F = \sum_{n=0}^\infty (1+n)^{s/2}F_n.$$

It is easy to check that the following holds

$$V_{H,T}(I - L)^{s/2}F = (I - L_H)^{s/2}V_{H,T}F, \quad \forall\, F \in \mathbb{P}.$$

and

$$V_{H,T}^{-1}(I - L_H)^{s/2}F = (I - L)^{s/2}V_{H,T}^{-1}F, \quad \forall\, F \in \mathbb{P}.$$

The following theorem is an analogue of the Meyer's inequality for fBm.

THEOREM 6.17. *For all $1 < p < \infty$ and non-negative integer k there are positive constants $c_{k,p}$ and $C_{k,p}$ (independent of H) such that for all $F \in \mathbb{D}_{k,p}^H$,*

$$(6.26) \qquad c_{k,p}\|F\|_{H,k,p} \leq \|(I - L_H)^{k/2}F\|_p \leq C_{k,p}\|F\|_{H,k,p}\,.$$

Proof We use the probability structure preserving mapping technique. Let us prove the second inequality. First by the property of probability structure preserving mapping, we have

$$(6.27) \quad \mathbb{E}\left(\int_0^T \cdots \int_0^T |D_{t_1,\cdots,t_k}^k F|^2 dt_1 \cdots dt_k\right)^{p/2}$$

$$= \mathbb{E}\left(\int_0^T \cdots \int_0^T |V_{H,T} D_{t_1,\cdots,t_k}^k F|^2 dt_1 \cdots dt_k\right)^{p/2}$$

$$= \mathbb{E}\left(\int_0^T \cdots \int_0^T |\mathbb{I}_{H,T}^*(t_1)\cdots\mathbb{I}_{H,T}^*(t_k) D_{t_1,\cdots,t_k}^{H,k} V_{H,T}F|^2 dt_1 \cdots dt_k\right)^{p/2}$$

$$= \mathbb{E}\left(\|D^{H,k} V_{H,T} F\|_{\Theta_H^{\otimes k}}^p\right)\,.$$

This implies that

$$\|V_{H,T}F\|_{H,k,p} = \|F\|_{k,p} \quad \forall\, F \in \mathbb{D}_{k,p}$$

and

$$\|V_{H,T}^{-1}F\|_{k,p} = \|F\|_{H,k,p} \quad \forall\, F \in \mathbb{D}_{k,p}^H\,.$$

By the Meyer's inequality for sBm (see for example [84]) and [86]), we have

$$\begin{aligned}
\|(I - L_H)^{k/2}F\|_p &= \|V_{H,T}^{-1}(I - L_H)^{k/2}F\|_p = \|(I - L)^{k/2}V_{H,T}^{-1}F\|_p \\
&\leq C_{k,p}\|V_{H,T}^{-1}F\|_{k,p} \\
&= C_{k,p}\|F\|_{H,k,p}\,.
\end{aligned}$$

The other inequality in (6.26) may be proved in a similar way. $\square$

We are to introduce another differential operator. Let

$$G = g\left(\int_0^T \eta_1(t)dB_t^H, \cdots, \int_0^T \eta_k(t)dB_t^H\right)$$

be as before. We define

$$\mathbb{D}_s^H G = \left[\mathbf{B}_s^H D_s^H\right] G\,,$$

where $\mathbf{B}^H = \mathbb{I}_{H,T}\mathbb{I}_{H,T}^*$. From (5.30), we have

$$\begin{aligned}
\left[\mathbf{B}_s^H D_s^H\right] G &= Hs^{2H-1}D_0^H G + H|T - s|^{2H-1}D_T^H G \\
&\quad + H\int_0^T |s - u|^{2H-1}\text{sign}\,(s - u)\frac{d}{du}D_u^H G du\,.
\end{aligned}$$

REMARK 6.18. In what follows we shall use the above convention that $\mathbf{B}_s^H f(s)$ is a function obtained by applying $\mathbf{B}^H$ to f which is considered as a function of s. This convention applies even when f is a function of many other variables. For example, if f is a function of s and t, then

$$\begin{aligned}
\mathbf{B}_t^H f(s,t) &= Ht^{2H-1}f(s,0) + H|T - t|^{2H-1}f(s,T) \\
&\quad + H\int_0^T |t - u|^{2H-1}\text{sign}\,(t - u)\frac{d}{du}f(s,u)du\,.
\end{aligned}$$

EXAMPLE 6.19. For any $f \in \mathbf{S}$,

$$\mathbb{D}_t^H \int_0^T f(s) dB_s^H = \mathbf{B}_t^H f(t).$$

EXAMPLE 6.20. For $f \in \mathbf{S}$

$$\mathbb{D}_t^H \exp\left\{\int_0^T f(s) dB_s^H\right\} = \exp\left\{\int_0^T f(s) dB_s^H\right\} \mathbf{B}^H f(t).$$

EXAMPLE 6.21. More generally, if $F = I_n(f_n)$, where $f_n \in \mathbf{S}^{\otimes n}$, then

$$\mathbb{D}_t^H F = n I_{n-1}(\mathbf{B}_t^H f_n(t)),$$

where

$$\mathbf{B}_t^H f_n(t)(t_1, \cdots, t_{n-1}) = \mathbf{B}_t^H f_n(t, t_1, \cdots, t_{n-1}).$$

THEOREM 6.22. *Let* $F(t), t \in [0, T] \times (\Omega, \mathcal{F}, P^H) \to \mathbb{R}$ *be measurable and such that* $(F(t), 0 \leq t \leq T)$ *is integrable. Let* $\mathbb{D}_t^H F(t) \in L^2(\Omega, \mathcal{F}, P^H)$. *Then*

$$(6.28) \qquad \mathbb{D}_t^H \left(\int_0^T F(s) dB_s^H\right) = \int_0^T \mathbb{D}_t^H F(s) dB_s^H + \mathbf{B}_t^H F(t).$$

Proof By linearity and density argument, it suffices to show (6.28) for $F(t) = I_n(f_n(t))$, where $f_n(t) = f_n(t; t_1, \cdots, t_n), t_1, \cdots, t_n \in [0, T]$ in $\mathbf{S}^{\otimes n}$ for all $t \in [0, T]$ (as a function of $t_1, \cdots, t_n$) and $f_n \in \mathbf{S}_{n+1}$ as a function of $t, t_1, \cdots, t_n$. Denote

$$g_{n+1}(t_1, \cdots, t_{n+1}) = \mathrm{Sym}_{t_1, \cdots, t_{n+1}} f_n(t_{n+1}; t_1, \cdots, t_n).$$

Then from Example 6.22 the left hand side of Eq. (6.28) is

$$\begin{aligned} \mathrm{LHS} &= \mathbb{D}_t^H (I_{n+1}(g_{n+1})) \\ &= (n+1) I_n(\mathbf{B}_t^H g_{n+1}(t, \cdot)). \end{aligned}$$

However,

$$\begin{aligned} &\mathbf{B}_t^H g_{n+1}(t, t_1, \cdots, t_n) \\ &= \frac{1}{n+1} \sum_{i=1}^{n+1} \mathbf{B}_t^H f_n(t_i; t_1, \cdots, t_{i-1}, t_{i+1}, \cdots, t_n, t) \\ &= \frac{1}{n+1} \mathbf{B}_t^H f_n(t; t_1, \cdots, t_n) + \frac{1}{n+1} \sum_{i=1}^{n} \mathbf{B}_t^H f_n(t_i; t_1, \cdots, t_{i-1}, t_{i+1}, \cdots, t_n, t) \end{aligned}$$

Thus

$$\begin{aligned} \mathrm{LHS} &= I_n(\mathbf{B}_t^H f_n(t; \cdot)) + \sum_{i=1}^{n} I_n(\mathbf{B}_t^H f_n(t_i; t_1, \cdots, t_{i-1}, t_{i+1}, \cdots, t_n, t)) \\ &= \mathbf{B}_t^H I_n(f_n(t; \cdot)) + \sum_{i=1}^{n} I_n(\mathbf{B}_t^H f_n(t_i; t_1, \cdots, t_{i-1}, t_{i+1}, \cdots, t_n, t)) \\ (6.29) \qquad &= \mathbf{B}_t^H F(t) + \sum_{i=1}^{n} I_n(\mathbf{B}_t^H f_n(t_i; t_1, \cdots, t_{i-1}, t_{i+1}, \cdots, t_n, t)). \end{aligned}$$

On the other hand, we see easily that

$$\int_0^T \mathbb{D}_t^H F(s) dB_s^H = \int_0^T n I_{n-1}(\mathbf{B}_t^H f_n(s; t_1, \cdots, t_{n-1}, t) dB_s^H$$

$$(6.30) \qquad = \sum_{i=1}^n \int_{\mathbb{R}^n} \mathbf{B}_t^H f_n(t_i; t_1, \cdots, t_{i-1}, t_{i+1}, \cdots, t_n, t) dB_{t_1}^H \cdots dB_{t_n}^H.$$

Combining (6.29) and (6.30), we obtain (6.28). $\qquad\qquad\qquad\qquad\qquad\square$

The following theorem states that the stochastic integral $\int_0^T f(t) dB_t^H$ that we introduced early is the divergence operator corresponding to $\mathbb{D}_t^H$.

THEOREM 6.23. *Let $f : [0,T] \otimes (\Omega, \mathcal{F}, P^H) \to \mathbb{R}$ be jointly measurable and let F be integrable in the sense of Definition of 6.11 and let $\int_0^T f(t) dB_t^H$ be an element of $L^2(\Omega, \mathcal{F}, P^H)$. Let $G \in D_{1,2}^H$ and $D_s^H G$ be in the domain of $\mathbb{I}_{H,T} \mathbb{I}_{H,T}^*$ such that $\mathbb{I}_{H,T}(s) \mathbb{I}_{H,T}^*(s) D_s^H G$ is in L^2. Then*

$$(6.31) \qquad \mathbb{E}\left(\int_0^T f(t) dB_t^H G \right) = \int_0^T \mathbb{E}\left(f(t) \mathbb{D}_t^H G \right) dt.$$

Proof Let $F(t)$ and G satisfy the above conditions. Then from the definition 6.11, the property of probability structure preserving mapping, and the property of stochastic integral for sBm it follows that

$$\mathbb{E}\left[\int_0^T f(t) dB_t^H G \right] = \mathbb{E}\left[V_{H,T}^{-1} \left(\int_0^T f(t) dB_t^H \right) V_{H,T}^{-1} G \right]$$

$$= \mathbb{E}\left[\int_0^T \mathbb{I}_H^* V_{H,T}^{-1} f(t) dB_t V_{H,T}^{-1} G \right]$$

$$= \mathbb{E}\left[\int_0^T \mathbb{I}_H^* V_{H,T}^{-1} f(t) D_t V_{H,T}^{-1} G dt \right]$$

From Lemma 6.15 it follows that

$$\mathbb{E}\left[\int_0^T f(t) dB_t^H G \right] = \mathbb{E}\int_0^T \mathbb{I}_{H,T}^*(t) V_{H,T}^{-1} f(t) \left[V_{H,T}^{-1} \mathbb{I}_H^*(t) D_t^H G \right] dt$$

$$= \mathbb{E}\left[\int_0^T V_{H,T} \mathbb{I}_H^*(t) V_{H,T}^{-1} f(t) \left[\mathbb{I}_{H,T}^*(t) D_t^H G \right] dt \right]$$

$$= \mathbb{E}\left[\int_0^T \mathbb{I}_H^*(t) f(t) \left[\mathbb{I}_H^*(t) D_t^H G \right] dt \right]$$

$$= \mathbb{E}\left[\int_0^T f(t) \mathbb{I}_H \mathbb{I}_H^* D_t^H G dt \right] = \mathbb{E}\left[\int_0^T f(t) \mathbb{D}_t^H G dt \right],$$

proving the theorem. $\qquad\qquad\qquad\qquad\qquad\qquad\qquad\qquad\qquad\qquad\square$

From the above proof we have also

$$(6.32) \qquad \mathbb{E}\left[\int_0^T f(t) dB_t^H G \right] = \mathbb{E}\int_0^T \mathbb{I}_{H,T}^*(t) f(t) \mathbb{I}_{H,T}^*(t) D_t^H G dt.$$

Denote by $\mathbb{L}_H^{1,2} = \mathbb{L}^{1,2}(\Omega, \mathcal{F}, P^H)$ the space of all $f : (\Omega, \mathcal{F}, P^H) \to \Theta_H([0,T])$ such that

$$\mathbb{E}\|f\|_{\Theta_H}^2 + \mathbb{E}\left\{\int_0^T \int_0^T |\mathbb{D}_s^H u(t)|^2 ds dt\right\} < \infty.$$

PROPOSITION 6.24. Let $F = (f(t), t \in [0,T])$ and $G = (g(t), t \in [0,T])$ be in $\mathbb{L}^2(\Omega, \mathcal{F}, P^H)$. Then

$$(6.33) \qquad \mathbb{E}\left[\int_0^T f(t)dB_t^H \int_0^T g(t)dB_t^H\right]$$

$$= \int_0^T \mathbb{E}\left[\mathbb{I}_{H,T}^* f(t) \mathbb{I}_{H,T}^* g(t)\right] dt + \int_0^T \int_0^T \mathbb{E}\left[\mathbb{D}_s^H f(t)\mathbb{D}_t^H g(s)\right] ds dt$$

$$= \int_0^T \mathbb{E}\left\{\left[\mathbf{B}^H f(t)\right] g(t)\right\} dt + \int_0^T \int_0^T \mathbb{E}\left[\mathbb{D}_s^H f(t)\mathbb{D}_t^H g(s)\right] ds dt.$$

Proof We shall use repeatedly Equations (6.31) and (6.32).

$$\mathbb{E}\left[\int_0^T f(t)dB_t^H \int_0^T g(t)dB_t^H\right]$$

$$= \int_0^T \mathbb{E}\left[f(t)\mathbb{D}_t^H \left(\int_0^T g(t)dB_t^H\right)\right] dt$$

$$= \int_0^T \mathbb{E}\left[f(t)\mathbb{I}_{H,T}(t)\mathbb{I}_{H,T}^*(t)g(t)\right] dt + \int_0^T \mathbb{E}\left[f(t)\int_0^T \mathbb{D}_t^H g(s)dB_s^H\right] dt$$

$$= \int_0^T \mathbb{E}\left[\mathbb{I}_{H,T}^* f(t)\mathbb{I}_{H,T}^* g(t)\right] dt + \int_0^T \int_0^T \mathbb{E}\left[\mathbb{D}_s^H f(t)\mathbb{D}_t^H g(s)\right] ds dt.$$

This is the right hand side of Eq. (6.33). $\qquad \square$

In particular we have

PROPOSITION 6.25. Let $f : (\Omega, \mathcal{F}, P^H) \to \Theta_H$ be in $\mathbb{L}^2(\Omega, \mathcal{F}, P^H)$. Then $\int_0^T f(t)dB_t^H$ exists in $L^2(\Omega, \mathcal{F}, P^H)$ and

$$(6.34) \qquad \mathbb{E}\left(\int_0^T f(t)dB_t^H\right) = 0$$

and

$$(6.35) \qquad \mathbb{E}\left(\int_0^T f(t)dB_t^H\right)^2 = \mathbb{E}\left(\|f\|_{\Theta_H}^2\right) + \mathbb{E}\int_0^T \int_0^T \mathbb{D}_s^H f(t)\mathbb{D}_t^H f(s)ds dt.$$

REMARK 6.26. When $H > 1/2$, a similar formula to (6.35) is obtained in [**35**] (see p. 591, Equation 3.17). But there is an error in that formula. Equation 3.17 in [**35**] should read as (6.35). The source of this error comes from Corollary 3.8 of [**35**]. The identity

$$\mathbb{E}\left(\left[F \diamond \int_0^\infty g_s dB_s^H\right]\left[G \diamond \int_0^\infty h_s dB_s^H\right]\right) = \mathbb{E}\left[D_{\Phi g}F D_{\Phi h}G + FG\langle g,h\rangle_\phi\right]$$

in Corollary 3.8 of [**35**] should be changed to

$$\mathbb{E}\left(\left[F \diamond \int_0^\infty g_s dB_s^H\right]\left[G \diamond \int_0^\infty h_s dB_s^H\right]\right) = \mathbb{E}\left[D_{\Phi h}F D_{\Phi g}G + FG\langle g,h\rangle_\phi\right].$$

6.5. Note on Stochastic Integral for fBm

There have been considerable amount of papers on the stochastic integral theory for fractional Brownian motion or for general Gaussian processes. In this subsection I will present a brief review. Since this theory is still in development it is impossible to give a complete list of references. I thank the referee for bringing me the attention of the papers [4], [20], [63], [93].

The stochastic integral for general Gaussian processes dates back at least to the paper [18]. Stochastic integral specifically for fractional Brownian motion of Hurst parameter $H > 1/2$ has been dealt with in [25], [67], and [74]. In those papers the stochastic integral, denoted by $\int_0^T f(s)\delta B_s^H$, is of *pathwise type* for the following reasons. First, the expectation of such integral is **not** always zero, *i.e.* $\mathbb{E}\left[\int_0^T f(s)\delta B_s^H\right]$ may not be zero for some integrand f (see [35] for examples). Secondly, the chain rule associated with this integral is similar to the ordinary one. Roughly speaking we have

$$\delta g\left(\int_0^t f(s)\delta B_s^H\right) = g'\left(\int_0^t f(s)\delta B_s^H\right) f(s)\delta B_s^H.$$

The fascinating Itô formula does not appear for this type of stochastic integral. However, if we use stochastic differential system $\dot{x}_t = b(x_t) + \sigma(x_t)\dot{B}_t^H$ as a mathematical model, then the ("drift") term $b(x_t)$ usually contains all the mean rate of change of the state x_t, and $\sigma(x_t)\dot{B}_t^H$ is the random fluctuation. It is important to assume that the mean rate change is contained in $b(x_t)$ and then then random fluctuation part $\sigma(x_t)\dot{B}_t^H$ has no more mean contribution. This motivates us to introduce a mean zero stochastic integral. This cannot be done by analogous argument of (1.2)-(1.3). As shown in [35], for fractional Brownian motion of Hurst parameter $H > 1/2$, (1.2)-(1.3) yield the same limit. In [35] a different stochastic integral is defined by replacing the product in (1.2) by the Wick product. This stochastic integral has mean zero property: If this stochastic integral is denoted by $\int_0^T f(t)dB_t^H$, then $\mathbb{E}\left(\int_0^T f(t)dB_t^H\right) = 0$. Moreover, we also obtain an Itô formula (see (11.9) below). In a particular case, we have

$$F(B_t^H) = F(B_s^H) + \int_s^t F'(B_r)dB_r^H + H\int_s^t r^{2H-1}F''(B_r)dr.$$

It reduces to the well-known Itô formula when $H = 1/2$. Because of this property we call this integral, $\int_0^T f(t)dB_t^H$, of Itô stochastic integral.

In [35] a fundamental tool used is the Malliavin calculus. In fact the Malliavin calculus is also used in another interesting and important paper [33]. The paper [33] is completed about the same time or a little bit early than [35]. However, these two papers are completed independently. In fact, in the same time, other relevant papers [27], [28] are also completed independently. The stochastic integral in [33] and [35] are different. This can be seen from the fact that Theorem 4.8 of [33]) states that if u is an adapted process, then $\int_0^t u_s \delta_H B_s^H = \int_0^t u_s dB_s$, where the later one is the Itô integral with respect to the standard Brownian motion. Thus if $u_s \equiv 1$, $\int_0^t 1\delta_H B_s^H = B_t$ is the standard Brownian motion. In our paper we have $\int_0^t 1\delta_H B_s^H = B_t^H$, not B_t!

Now there have been considerable amount of papers on the stochastic calculus of fractional Brownian motion. Some works extend pathwise stochastic integral to more general case. In [37], [42], [43] an Itô-Stratonovich formula (the ordinary chain rule) is obtained for fBm with Hurst parameter $H > 1/6$ (see also [21]). There is a milestone work of T. Lyons which develops an integration theory for general rough path by using the K.T. Chen's iterated integrals [78] and [79]. This theory can be applied to study the stochastic integral for fractional Brownian motion in [24].

A stochastic integral of Itô type with respect to fractional Brownian motion of Hurst parameter $H < 1/2$ is defined in [2] by using approximation. More precisely, they define stochastic integral with respect to

$$\tilde{B}_t^H = \int_0^t (t - s)^{H - \frac{1}{2}} dB_s ,$$

which is the main part of fBm. In fact they define

$$\int_0^T \phi(t) d\tilde{B}_t^H = \lim_{\varepsilon \to 0} \int_0^T \phi(t) d\tilde{B}_t^{H,\varepsilon} ,$$

where $\tilde{B}_t^{H,\varepsilon} = \int_0^t (t - s + \varepsilon)^{H - \frac{1}{2}} dB_s$, $\varepsilon > 0$, is a semimartingale. Recall the Riemann-Liouville integral operator

$$I_{T-}^\alpha f(s) = \frac{(-)^\alpha}{\Gamma(\alpha)} \int_s^T (r - s)^{\alpha - 1} f(r) dr .$$

It is proved in [2] that if ϕ is a process satisfying the following condition

$$\phi \in I_{T-}^{H - \frac{1}{2}}(\mathbb{L}^{1,2})$$

and

$$\int_0^T \int_0^r |D_s \phi_r| (r - s)^{H - \frac{3}{2}} ds dr < \infty ,$$

then $\int_0^T \phi(r) d\tilde{B}_r$ exists. However, this condition is not satisfied for $\phi(r) = \tilde{B}_r^H$ if $H < 1/2$. In fact, we have

$$D_s \tilde{B}_r^H = \chi_{[0,r]}(s)(r - s)^{H - \frac{1}{2}} .$$

Thus

$$\int_0^r |D_s \phi_r| (r - s)^{H - \frac{3}{2}} ds = \int_0^r (r - s)^{2H - 2} ds$$

which is finite if and only if $2H - 2 > -1$ or $H > 1/2$. Thus under this framework $\int_0^T B_s^H dB_s^H$ is not well-defined for $H < 1/2$.

In [3] a stochastic integral is defined for general Gaussian process and an Itô type formula is established. This stochastic integral applies to fractional Brownian motion of Hurst parameter $H > 1/4$ and the Itô formula appears to be different than the one in [35] when applied to fractional Brownian motion for $H > 1/2$. In Chapter 11 we shall recall these two Itô formulas and illustrate that they coincide.

In [58] a stochastic calculus is developed in the framework of white noise analysis and some results such as Girsanov formula (of constant translation), Clark-Haussman-Ocone formula are established and applied to the option pricing theory of fractal market. Along this framework an Itô formula is proved in [4] and [14] for all Hurst parameter H. A similar formula is proved in Chapter 11 (Theorem 11.1) of this paper independently and within the framework of this paper. Our formula holds for more general processes than fractional Brownian motions.

[**31**] deals with the stochastic analysis for the so-called Volterra process which is special Gaussian process. [**63**] considers the stochastic integral for the so-called Volterra processes which admits a representation of Volterra type with respect to standard Brownian motion. As stated in that paper, this is an extension of the work [**33**]. Approximation based on the famous Karhunen-Loéve expansion is also discussed.

In [**93**] a stochastic integral is introduced for a very general class of processes and an Itô formula is also obtained. But the theory is complicate and it seems simpler to establish the integral theory for fractional Brownian motion directly.

Very recently in [**21**] the stochastic integral for fractional Brownian motion of any Hurst parameter is defined for a larger class of integrands.

CHAPTER 7

Stochastic Integration

7.1. Existence and Examples

We have defined stochastic integral in Chapter 6.3 for general integrands and for any Hurst parameter H. Now we study some properties of this stochastic integral. The following theorem is a direct consequence of Proposition 6.25.

PROPOSITION 7.1. (a) Let $f : (\Omega, \mathcal{F}, P^H) \to \Theta_H([0,T])$ be measurable (formally we may write f as $f(t, \omega)$ although it might be a generalized function of t) and

$$(7.1) \qquad \mathbb{E} \, \|f(\cdot)\|^2_{\Theta_H} < \infty \,,$$

where

$$(7.2) \qquad \|f(\cdot)\|^2_{\Theta_H} = \int_0^T |\, (\mathbb{I}^*_H f)\,(t)|^2 dt \,.$$

(b) As a function of s and t, $D_t^H f(s)$ exists as an element of $\Theta_H^{\otimes 2}$ for almost every $\omega \in \Omega$ and

$$(7.3) \qquad \mathbb{E} \int_0^T \int_0^T \left[(\mathbb{I}^*_{H,T})^{\otimes 2} D_s^H f(t) (\mathbb{I}^*_{H,T})^{\otimes 2} D_t^H f(s) \right] ds dt < \infty \,,$$

where if for P^H- almost all $\omega \in \Omega$, $D_s^H f(t)$ is continuously differentiable function of s and t, then

$$(7.4)$$

$$\left((\mathbb{I}^*_{H,T})^{\otimes 2} \right) D_s^H f(t) = \kappa_H^2 (st)^{\frac{1}{2}-H} \frac{\partial^2}{\partial s \partial t} \int_s^T \int_t^T (uv)^{H-\frac{1}{2}} [(u-s)(v-t)]^{H-\frac{1}{2}} D_u^H f_v du dv \,.$$

Then $\int_0^T f(s) dB_s^H$ exists in L^2 and

$$(7.5) \qquad \mathbb{E} \left[\int_0^T f(s) dB_s^H \right] = 0$$

and

$$(7.6) \qquad \mathbb{E} \left[\int_0^T f(s) dB_s^H \right]^2 = \mathbb{E} \left\{ \|f(\cdot)\|^2_{\Theta_H} + \int_0^T \int_0^T \left[(\mathbb{I}^*_{H,T})^{\otimes 2} D_s^H f(t) \right] \left[(\mathbb{I}^*_{H,T})^{\otimes 2} D_t^H f(s) \right] ds dt \right\} .$$

Proof It follows easily from Proposition 6.24 (Equation (6.33)). □

The definition (6.11) of stochastic integral introduced in Chapter 6 is general. But as in the usual case it is always difficult to evaluate a stochastic integral. We give another definition which uses chaos expansion and an idea of creation operator from quantum field theory (see [**7**], [**83**] and the references therein).

Let $F : [0, T] \times (\Omega, \mathcal{F}, P^H) \to \mathbb{R}$ be given as

$$(7.7) \qquad F(t) = I_n^{H,T}(f_n(t; \cdot)) = \int_{[0,T]^n} f_n(t; t_1, \cdots, t_n) dB_{t_1}^H \cdots dB_{t_n}^H,$$

where $f_n(t; t_1, \cdots, t_n)$, $t, t_1, \cdots, t_n \in [0, T]$, is measurable with respect to $t, t_1, \cdots, t_n$ and is symmetric with respect to $t_1, \cdots, t_n$ for all $t \in [0, T]$. We assume that for each $t \in [0, T]$, $f_n(t; \cdot, \cdots, \cdot)$ is an element in $\Theta_H^{\otimes n}$. Let us define

$$
\begin{aligned}
\tilde{f}_{n+1}(t_1, \cdots, t_n, t_{n+1}) &= \mathrm{Sym}_{t_1, \cdots, t_{n+1}} f_n(t_{n+1}; t_1, \cdots, t_n) \\
(7.8) \qquad\qquad &= \frac{1}{n+1} \sum_{k=1}^{n+1} f_n(t_k; t_1, \cdots, t_{k-1}, t_{k+1}, \cdots, t_n).
\end{aligned}
$$

If $\tilde{f}_{n+1} \in \Theta_H^{\otimes(n+1)}$, then we say that F is *algebraically integrable* and define
$$(7.9)$$
$$\int_0^T F(t) dB_t^H = I_{n+1}^{H,T}(\tilde{f}_{n+1}) = \int_{[0,T]^{n+1}} \tilde{f}_{n+1}(t_1, \cdots, t_n, t_{n+1}) dB_{t_1}^H \cdots dB_{t_{n+1}}^H.$$

Now let be given $F(t) \in L^2(\Omega, \mathcal{F}, P^H)$, $t \in [0, T]$. Thus

$$(7.10) \qquad\qquad F(t) = \sum_{n=0}^{\infty} F_n(t) = \sum_{n=0}^{\infty} I_n^{H,T}(f_n(t)).$$

Assume that $f_n(t; t_1, \cdots, t_n)$, $t, t_1, \cdots, t_n \in [0, T]$, is measurable with respect to $t, t_1, \cdots, t_n$ and is symmetric with respect to $t_1, \cdots, t_n$ for all $t \in [0, T]$. Define

$$\int_0^T F_n(t) dB_t^H := I_{n+1}^{H,T}(\tilde{f}_{n+1})$$

by the above approach.

DEFINITION 7.2. Let $(F(t), 0 \le t \le T)$ be given by (7.10) and $F_n(t)$ is algebraically integrable. If $\sum_{n=0}^{\infty} \int_0^T F_n(t) dB_t^H$ is convergent in $L^2(\Omega, \mathcal{F}, P^H)$, then we say that F is algebraically integrable and we denote the above sum by

$$\int_0^T F(t) dB_t^H = \lim_{k \to \infty} \sum_{n=0}^{k} \int_0^T F_n(t) dB_t^H.$$

REMARK 7.3. (a)In the above definition one may also require that the L^2 convergence be replaced by the convergence in probability or other types of convergence.

(b) This definition of stochastic integral appeared in a more general form in the literature of quantum field theory or quantum probability within the framework of Fock space (see for instance [**40**], [**83**] and the references therein).

(c) In the expansion f_n can be also generalized function. But we shall not discuss this general case in the following sections.

THEOREM 7.4. *If F_t is algebraically integrable in the sense of Definition 7.2, then it is integrable in the sense of Definition 6.11. Moreover, both integrals coincide.*

Proof Assume that $F(t)$ is integrable in the sense of Definition 7.2. Without loss of generality, we may assume that $F(t)$ is of the form (7.7). Since $f_n(t; \cdot, \cdots, \cdot)$ is an element of $\Theta_H^{\otimes n}$ we have

$$V_{H,T}^{-1} F(t) = I_n((\mathbb{I}_H^*)^{\otimes n} f_n(t; \cdots)).$$

Thus

$$\mathbb{I}_{H,T}^* V_{H,T}^{-1} F(t) = I_n(\mathbb{I}_{H,T}^* \otimes (\mathbb{I}_{H,T}^*)^{\otimes n} f_n(\cdot; \cdots)).$$

It is obvious that $\mathbb{I}_{H,T}^* V_{H,T}^{-1} F$ is integrable and $\int_0^T \mathbb{I}_{H,T}^* V_{H,T}^{-1} F(t) dB_t = I_{n+1}(\tilde{g}_{n+1})$, where $\tilde{g}_{n+1}$ is the symmetrization of $\mathbb{I}_{H,T}^* \otimes (\mathbb{I}_{H,T}^*)^{\otimes n} \tilde{f}_{n+1}(\cdot; \cdots)$. Therefore $(F(t), 0 \le t \le T)$ is also integrable in the sense of Definition (6.11). Now it is easy to show that the symmetrization of $\mathbb{I}_{H,T}^* \otimes (\mathbb{I}_{H,T}^*)^{\otimes n} f_n(\cdot; \cdots)$ is the same as $(\mathbb{I}_{H,T}^*)^{\otimes (n+1)} \tilde{f}_{n+1}$. Thus

$$\begin{aligned}
\int_0^T F(t) dB_t^H &= V_{H,T} \left[I_{n+1}((\mathbb{I}_{H,T}^*)^{\otimes (n+1)} \tilde{f}_{n+1}) \right] \\
&= I_{n+1}^{H,T}(\tilde{f}_{n+1}).
\end{aligned}$$

This proves the theorem. $\qquad\square$

It is clear that

$$B_t^H = I_1(\chi_{[0,t]}(\cdot)).$$

Let $g(t, t_1) = \chi_{[0,T]}(t) \chi_{[0,t]}(t_1)$. Then

$$\begin{aligned}
\tilde{g}(t_1, t_2) &= \mathrm{Sym}_{t_1, t_2} \left(\chi_{[0,T]}(t_2) \chi_{[0,t_2]}(t_1) \right) \\
&= \frac{1}{2} \chi_{[0,T]}(t_1) \chi_{[0,T]}(t_2).
\end{aligned}$$

Therefore it is easy to see that $\tilde{g} \in \Theta_H^{\otimes 2}$. From (6.8) and (7.9), it follows that

$$\begin{aligned}
\int_0^T B_t^H dB_t^H &= I_2^{H,T}(\tilde{g}) \\
&= \frac{1}{2} \left[\left(\int_0^T \chi_{[0,T]}(t) dB_t^H \right)^2 - \|\chi_{[0,T]}\|_{\Theta_H}^2 \right] \\
&= \frac{1}{2} \left[\left(\int_0^T \chi_{[0,T]}(t) dB_t^H \right)^2 - \int_0^T |\mathbb{I}_{H,T}^* \chi_{[0,T]}(t)|^2 dt \right]
\end{aligned}$$

From the expression of $\mathbb{I}_{H,T}^*$, we see that

$$\mathbb{I}_{H,T}^* \chi_{[0,T]} = Z_H(T, t).$$

Thus by (2.15) we obtain

$$\|\chi_{[0,T]}\|_{\Theta_H}^2 = \int_0^T Z_H(T, t)^2 dt = T^{2H}.$$

Therefore

$$(7.11) \qquad \int_0^T B_t^H dB_t^H = \frac{1}{2} \left[\left(B_T^H \right)^2 - T^{2H} \right].$$

We state the above result as

EXAMPLE 7.5. Let $0 < H < 1$. Then $\int_0^T B_t^H dB_t^H$ is well-defined and

$$(7.12) \qquad \left(B_T^H\right)^2 = 2 \int_0^T B_t^H dB_t^H - T^{2H} .$$

EXAMPLE 7.6. Let $0 < H < 1$ and let $f \in \Theta_H \cap L^2([0,T])$ and let $F(t) = f(t)I_n(f^{\otimes n})$, $t \in [0,T]$. It is easy to see that

$$\int_0^T F(t)dB_t^H = I_{n+1}(f^{\otimes(n+1)})$$

$$(7.13) \qquad = \sum_{k \leq (n+1)/2} \frac{(-1)^k (n+1)! \|f\|_{\Theta_{H,T}}^{2k}}{2^k k! (n+1-2k)!} \left(\int_0^T f(t)dB_t^H \right)^{n+1-2k} .$$

EXAMPLE 7.7. Let $0 < H < 1$ and let $f \in \Theta_H \cap L^2([0,T])$. Denote

$$(7.14) \qquad \mathbf{E}\left(\int_0^T f(t)dB_t^H \right) = \exp\left\{ \int_0^T f(t)dB_t^H - \frac{1}{2}\|f\|_{\Theta_H}^2 \right\} .$$

Now we want to show that for any complex number u, $\mathbf{E}\left(uB_t^H\right)$, $t \in [0,T]$, is integrable and

$$(7.15) \qquad \mathbf{E}(uB_T^H) = 1 + u \int_0^T \mathbf{E}\left(uB_t^H\right) dB_t^H .$$

Let u be a complex number. First it is easy to see that

$$\mathbf{E}\left(uB_t^H\right) = \sum_{n=0}^{\infty} \frac{u^n}{n!} I_n^{H,T}(f_n(t)) ,$$

where

$$(7.16) \qquad f_n(t; t_1, \cdots, t_n) = \chi_{[0,t]}(t_1) \cdots \chi_{[0,t]}(t_n) .$$

Define

$$g_{n+1}(t_1, \cdots, t_{n+1}) := \chi_{[0,T]}(t_{n+1})\chi_{[0,t_{n+1}]}(t_1) \cdots \chi_{[0,t_{n+1}]}(t_n) , \quad t_1, \cdots, t_{n+1} \in [0,T] .$$

It is easy to check that

$$\tilde{g}_{n+1}(t_1, \cdots, t_{n+1}) = \mathrm{Sym}_{t_1, \cdots, t_{n+1}} g_{n+1}(t_1, \cdots, t_{n+1})$$

$$= \frac{1}{n+1}\chi_{[0,T]}(t_1) \cdots \chi_{[0,T]}(t_{n+1}) .$$

Therefore

$$\int_0^T I_n^H(\chi_{[0,t]}^{\otimes n})dB_t^H = \frac{1}{n+1}I_{n+1}^H\left(\chi_{[0,T]}^{\otimes(n+1)}\right) .$$

From this it is easy to show that $\mathbf{E}\left(uB_t^H\right)$ is integrable and

$$\int_0^T \mathbf{E}\left(uB_t^H\right) dB_t^H = \sum_{n=0}^{\infty} \frac{u^n}{(n+1)!} I_{n+1}^H\left(\chi_{[0,T]}^{\otimes(n+1)}\right)$$

$$= \frac{1}{u} \sum_{n=1}^{\infty} \frac{u^n}{n!} I_n^H\left(\chi_{[0,T]}^{\otimes(n)}\right)$$

$$(7.17) \qquad = \frac{1}{u}\left[\mathbf{E}\left(uB_t^H\right) - 1 \right] .$$

7.2. Stochastic Integral $\int_0^a f(t)dB_t^H$ of different upper limits

Let $0 < a < T$ be given. Consider the stochastic integral $\int_0^a g(t)dB_t^H$, where g is a stochastic process such that it is continuously differentiable with respect to t for almost all $\omega \in \Omega$. This stochastic integral can be considered as $\int_0^T g(t)\chi_{[0,a]}(t)dB_t^H$ and is well-defined by Definition 6.11. However, we may also define the stochastic integral using the same definition but by replacing the T by a in the definition 6.11. The integral obtained this way is called a-integral temporarily. The $\mathbb{T}_{H,a}$ and $V_{H,a}$ are different in these two different definitions. So we may obtain two integrals. We shall illustrate that these two stochastic integrals are the same. First let us consider the case that g is deterministic. Let $h(t) = g(t)\chi_{[0,a]}$ with $0 < a < T$. From Lemma 5.9 it follows that when $0 \le s \le a$,

$$
\begin{aligned}
\mathbb{T}_{H,T}^H h(s) &= \kappa_H s^{\frac{1}{2}-H} \frac{d}{ds} \int_s^T t^{H-\frac{1}{2}}(t-s)^{H-\frac{1}{2}} g(t)\chi_{[0,a]}(t)dt \\
&= \begin{cases} \mathbb{T}_{H,a}^* g(s) & if \ \ 0 \le s \le a \\ 0 & if \ \ a < s \le T \end{cases}.
\end{aligned}
$$

Therefore

$$
\int_0^a g(t)dB_t^H = \int_0^T g(t)\chi_{[0,a]}(t)dB_t^H ,
$$

where the second integral is defined by using Definition 6.11 and the first integral is the a-integral with T replaced by a.

Let f_n be smooth function of n variables with all bounded derivatives. Assume that the support of f_n is in $[0,a]^n$. In this case it is also natural to define the multiple stochastic integral by using $\mathbb{T}_{H,a}^*$, denoted by $I_n^{H,a}(f_n)$. It is interesting to know if $I_n^{H,a}(f_n) = I_n^{H,T}(f_n)$. In fact we have

PROPOSITION 7.8. Let f_n be smooth function of n variables with the support being in $[0,a]^n$. Then

(7.18)
$$
\begin{aligned}
V_{H,T}\left[I_n(f_n)\right] &= V_{H,a}\left[I_n(f_n)\right] \\
&= I_n\left[\left(\mathbb{T}_{H,a}^T\right)^{\otimes n} f_n\right].
\end{aligned}
$$

If f is continuously differentiable with support in $[0,a]$, then

$$
V_{H,T}\left[\int_0^a f(t)dB_t\right] = V_{H,a}\left[\int_0^a f(t)dB_t\right].
$$

PROPOSITION 7.9. If f is continuously differentiable with support in $[0,a]$, then

(7.19)
$$
\int_0^a f(t)dB_t^H = \int_0^T f(t)\chi_{[0,a]}dB_t^H ,
$$

where the stochastic integral on the right hand side is defined by Definition 6.11 and the stochastic integral on the left hand side is defined by Definition 6.11 with T replaced by a.

7.3. An L_p estimate

In the sBm case the following inequality is a consequence of Meyer's inequality:

THEOREM 7.10. *Let $(f(t), 0 \leq t \leq T)$ be a stochastic process which is in* $\mathbb{L}^{1,2}(\Omega, \mathcal{F}, P^H)$. *Then for any $p \in (1, \infty)$ there is a constant C_p such that*

$$
\mathbb{E} \left| \int_0^T f(t) dB_t^H \right|^p \leq C_p \left\{ \mathbb{E} \left(\int_0^T |\mathbb{I}_{H,T}^* f(t)|^2 dt \right)^{p/2} \right.
$$

$$
(7.20) \qquad\qquad \left. + \mathbb{E} \left(\int_0^T \int_0^T |\mathbb{D}_s^H f(t)|^2 ds dt \right)^{p/2} \right\}.
$$

Proof Instead of following the routine proof from the Meyer's inequality, we use the technique of probability structure preserving mapping. It is well-known that in the sBm case the following inequality holds (see [86], Eq. (3.15)):
(7.21)

$$
\mathbb{E} \left(\left| \int_0^T u_t dB_t \right|^p \right) \leq C_p \left\{ \mathbb{E} \left(\int_0^T |u_t|^2 dt \right)^{p/2} + \mathbb{E} \left(\int_0^T \int_0^T |D_s u_t|^2 ds dt \right)^{p/2} \right\}
$$

for any $u \in \mathbb{L}^{1,2}(\Omega, \mathcal{F}, P)$. Now by definition 6.11 we have

$$
\mathbb{E} \left(\left| \int_0^T f(t) dB_t^H \right|^p \right) = \mathbb{E} \left| \int_0^T \mathbb{I}_{H,T}^* V_{H,T}^{-1} f(t) dB_t \right|^p
$$

$$
\leq C_p \left\{ \mathbb{E} \left(\int_0^T |\mathbb{I}_{H,T}^*(t) V_{H,T}^{-1} f(t)|^2 dt \right)^{p/2} \right.
$$

$$
\left. + \mathbb{E} \left(\int_0^T \int_0^T |D_s \mathbb{I}_{H,T}^*(t) V_{H,T}^{-1} f(t)|^2 ds dt \right)^{p/2} \right\}.
$$

From Lemma 6.15 and the property of probability preserving mapping it follows

$$
\mathbb{E} \left(\left| \int_0^T f(t) dB_t^H \right|^p \right) \leq C_p \left\{ \mathbb{E} \int_0^T |\mathbb{I}_{H,T}^* V_{H,T}^{-1} f(t)|^2 dt \right.
$$

$$
\left. + \mathbb{E} \left(\int_0^T \int_0^T V_{H,T}^{-1} \mathbb{I}_{H,T}^*(t) \mathbb{I}_{H,T}^*(s) D_s^H f(t) ds dt \right) \right\}
$$

$$
= C_p \left\{ \mathbb{E} \left(\int_0^T |\mathbb{I}_{H,T}^* f(t)|^2 dt \right)^{p/2} \right.
$$

$$
\left. + \mathbb{E} \left(\int_0^T \int_0^T |\mathbb{I}_{H,T}^*(t) \mathbb{I}_{H,T}^*(s) D_s^H f(t)|^2 ds dt \right)^{p/2} \right\}.
$$

This proves the theorem easily. $\square$

A direct consequence of this theorem is

COROLLARY 7.11. *Let f and f_n, $n = 1, 2, \cdots$ be jointly measurable as a mapping from $[0, T] \times (\Omega, \mathcal{F}, P^H)$ to $\mathbb{R}$. Suppose that f and $f_n, n = 1, 2, \cdots$ are elements* of $\mathbb{L}^{1,2}(\Omega, \mathcal{F}, P^H)$. If

$$\mathbb{E}\left(\int_0^T |\mathbb{I}^*_{H,T}(f_n(t) - f(t))|^2 dt\right)^{p/2}$$

$$+\mathbb{E}\left(\int_0^T \int_0^T |\mathbb{I}^*_{H,T}(t)\mathbb{I}^*_{H,T}(s)D^H_s(f_n(t) - f(t))|^2 ds dt\right)^{p/2} \to 0$$

as $n \to \infty$, then $\int_0^T f_n(t)dB^H_t$ converges to $\int_0^T f(t)dB^H_t$ in $L^p(\Omega, \mathcal{F}, P^H)$.

REMARK 7.12. Theorem 7.10 is a generalization of classical result (see e.g. [86]). We shall use this to study the continuity of integral process in Chapter 13.

7.4. An Example

Recall the following well-known identity

$$\mathbb{E}\left(e^X\right) = e^{\frac{1}{2}\mathbb{E}(X^2)}$$

for any centered (mean zero) Gaussian random variable X. Let x and y be any complex numbers and $0 \le s < t \le T$ and $0 \le u < v \le T$. It is clear that

$$\mathbb{E}\left\{e^{x(B^H_t - B^H_s) - \frac{x^2}{2}(t-s)^{2H}} e^{y(B^H_v - B^H_u) - \frac{y^2}{2}(v-u)^{2H}}\right\} = e^{\frac{xy}{2}\left[|t-u|^{2H} + |v-s|^{2H} - |t-v|^{2H} - |s-u|^{2H}\right]}.$$

On the other hand we have

$$e^{x(B^H_t - B^H_s) - \frac{x^2}{2}(t-s)^{2H}} = \sum_{n=0}^\infty \frac{x^n}{n!}(B^H_t - B^H_s)^{\diamond n}.$$

Therefore

$$(7.22) \quad \mathbb{E}\left[(B^H_t - B^H_s)^{\diamond m}(B^H_v - B^H_u)^{\diamond n}\right]$$
$$= \begin{cases} 0 & if \ m \ne n \\ \frac{n!}{2^n}\left[|t-u|^{2H} + |v-s|^{2H} - |t-v|^{2H} - |s-u|^{2H}\right]^n & if \ m = n. \end{cases}$$

Generally we have

$$\begin{cases} 0 & if \ m \ne n \\ \mathbb{E}\left[I_n(f_n)I_m(g_n)\right] = n!\langle f_n, g_n \rangle_{\Theta^{\otimes n}_H} & if \ m = n \end{cases}$$

Let us consider the stochastic integral $\int_0^T B^H_s dB^H_s$ by using the ordinary product. Consider a partition $\pi : 0 = t_0 < t_1 < \cdots t_{n-1} < t_n = T$. Let us consider the sum

$$(7.23) \qquad S_\pi := \sum_{k=0}^{n-1} B^H_{t_k}(B^H_{t_{k+1}} - B^H_{t_k}).$$

Note that in the above product the Wick product is **not** used. Taking expectation we have

$$
\begin{aligned}
\mathbb{E}\left(S_\pi\right) &= \sum_{k=0}^{n-1} \mathbb{E}\left[B_{t_k}^H (B_{t_{k+1}}^H - B_{t_k}^H)\right] \\
&= \sum_{k=0}^{n-1} \left\{ \frac{1}{2} \left[t_{k+1}^{2H} - t_k^{2H} - |t_{k+1} - t_k|^{2H} \right] \right\} \\
&= T^{2H} - \sum_{k=0}^{n-1} |t_{k+1} - t_k|^{2H} .
\end{aligned}
$$

The last term is divergent when $H < 1/2$.

Now we replace the ordinary product by the Wick product in the definition of S_π. The new sum, denoted by I_π, is given as follows

$$
(7.24) \qquad I_\pi := \sum_{k=0}^{n-1} B_{t_k}^H \diamond (B_{t_{k+1}}^H - B_{t_k}^H),
$$

where $\diamond$ denotes the Wick product (see for instance [35], [41], [46], [83], [98], and in particular the references therein). Then we will prove that I_π is convergent in $L^2(\Omega, \mathcal{F}, P^H)$ if and only if $H > 1/4$. It is easy to see that the expectation is zero. A simple algebra yields

$$
\begin{aligned}
2I_\pi &= 2\sum_{k=0}^{n-1} B_{t_k}^H \diamond (B_{t_{k+1}}^H - B_{t_k}^H) \\
&= \sum_{k=0}^{n-1} \left[\left(B_{t_{k+1}}^H\right)^{\diamond 2} - \left(B_{t_k}^H\right)^{\diamond 2} - (B_{t_{k+1}}^H - B_{t_k}^H)^{\diamond 2} \right] \\
&= (B_T^H)^{\diamond 2} - \sum_{k=0}^{n-1} (B_{t_{k+1}}^H - B_{t_k}^H)^{\diamond 2} .
\end{aligned}
$$

We shall find condition under which the above last sum, denoted by $\tilde{I}_\pi$, is convergent to 0 in L^2. It is obvious that

$$
\begin{aligned}
\mathbb{E}\left[\tilde{I}_\pi^2\right] &= \sum_{k,j=0}^{n-1} \mathbb{E}\left[(B_{t_{k+1}}^H - B_{t_k}^H)^{\diamond 2}\right]\left[(B_{t_{j+1}}^H - B_{t_j}^H)^{\diamond 2}\right] \\
&= 2\sum_{k>j} \mathbb{E}\left[(B_{t_{k+1}}^H - B_{t_k}^H)^{\diamond 2}\right]\left[(B_{t_{j+1}}^H - B_{t_j}^H)^{\diamond 2}\right] + \sum_{k=0}^{n-1} \mathbb{E}\left[(B_{t_{k+1}}^H - B_{t_k}^H)^{\diamond 2}\right]^2 \\
&=: I_{\pi,1} + I_{\pi,2} .
\end{aligned}
$$

By (7.22), it follows that

$$
I_{\pi,2} = 2\sum_{k=0}^{n-1} |t_{k+1} - t_k|^{4H}
$$

which is finite if and only if $4H \geq 1$. Now let us consider $I_{\pi,1}$. For simplicity, let us consider the uniform partition $t_k = kT/n$. Without loss of generality let us assume that $0 < H < 1/2$.

$$
\begin{aligned}
\frac{1}{2}I_{\pi,1} &= \sum_{k>j}\left[|t_{k+1}-t_j|^{2H}+|t_k-t_{j+1}|^{2H}-|t_{k+1}-t_{j+1}|^{2H}-|t_k-t_j|^{2H}\right]^2 \\
&= n^{-4H}\sum_{k>j}\left[|k-j+1|^{2H}+|k-j-1|^{2H}-2|k-j|^{2H}\right]^2 \\
&= n^{-4H}\sum_{l=1}^{n-1}\sum_{k=l+1}^{n-1}\left[|l+1|^{2H}+|l-1|^{2H}-2l^{2H}\right]^2 \\
&= 2n^{-4H}\sum_{l=1}^{n-1}(n-2-l)\left[|l+1|^{2H}+|l-1|^{2H}-2l^{2H}\right]^2 \\
&\leq Cn^{-4H+1}\sum_{l=1}^{n-1}\left[|l+1|^{2H}+|l-1|^{2H}-2l^{2H}\right]^2 \\
&\leq Cn^{-4H+1}.
\end{aligned}
$$

since when $l \to \infty$, $|l+1|^{2H}+|l-1|^{2H}-2l^{2H} \approx l^{2H-2}$. Thus, $I_{\pi,1}$ converges to 0 (as $n \to 0$) when $4H > 1$.

Summarizing the above computation we have

PROPOSITION 7.13. (a) Let $\pi : 0 = t_0 < t_1 < \cdots < t_n = T$ be a partition of $[0,T]$ and let

$$
S_\pi = \sum_{k=0}^{n-1} B_{t_k}^\pi\left(B_{t_{k+1}}^\pi - B_{t_k}^\pi\right).
$$

Then as $|\pi| \to 0$, S_π is a Cauchy sequence in $L^2(\Omega, \mathcal{F}, P^H)$ if and only if $H \geq 1/2$. In this case S_π converges to $\frac{1}{2}\left(B_T^H\right)^2$.

(b) Let $\pi : 0 = t_0 < t_1 < \cdots < t_n = T$ be a uniform partition of $[0,T]$ and let

$$
I_\pi = \sum_{k=0}^{n-1} B_{t_k}^\pi \diamond \left(B_{t_{k+1}}^\pi - B_{t_k}^\pi\right).
$$

Then as $|\pi| \to 0$, I_π is a Cauchy sequence in $L^2(\Omega, \mathcal{F}, P^H)$ if and only if $H \geq 1/4$. In this case I_π converges to $\frac{1}{2}\left(B_T^H\right)^2 - \frac{1}{2}T^{2H}$.

CHAPTER 8

Nonlinear Translation (Absolute Continuity)

Let Ω denote the space of all continuous functions on $[0, T]$ with the sup norm and let $\mathcal{F}$ be the Borel σ-algebra of Ω. Let P and P^H be the canonical Wiener and fractional Wiener measure on $(\Omega, \mathcal{F})$, respectively. We shall use $(\Omega, \mathcal{F}, P)$ to denote the probability space for sBm and $(\Omega^H, \mathcal{F}^H, P^H)$ to denote the probability space for fBm to emphasize that they are two different probability spaces. The natural filtrations are denoted by $\mathcal{F}_t^H = \mathcal{F}_t$, $0 \leq t \leq T$.

Assume that $f : [0, T] \times (\Omega^H, \mathcal{F}^H, P^H) \to \mathbb{R}$ is a (not necessarily) adapted process such that

$$\int_0^T |f(s, B_.^H)|^2 ds < \infty \quad \text{for almost all } \omega \in \Omega^H.$$

Consider an (anticipative) translation $\Lambda : (\Omega^H, \mathcal{F}^H, P^H) \to (\Omega^H, \mathcal{F}^H, P^H)$ of B^H

$$(8.1) \qquad \Lambda : \quad B_.^H \longrightarrow B_.^H + \int_0^{\cdot} f(s, B_.^H) ds.$$

Define an induced anticipative translation, $\Gamma : \Omega \to \Omega$, of B by

$$(8.2) \qquad \Gamma : \quad B_. \longrightarrow B_. + \int_0^{\cdot} g(s, B) ds,$$

where $g = \mathbb{B}_{H,T} V_{H,T}^{-1} f$. Note that even if f is $\mathcal{F}_t^H$-adapted, g may not be $\mathcal{F}_t$-adapted. From the computation of Chapter 5, f can be computed from g by $f - \mathbb{I}_{H,T} V_{H,T} g$. Let $\mathbf{E}$ denote the set of finite linear combinations of exponential functionals on $(\Omega, \mathcal{F}, P)$ and let $\tilde{\mathbf{E}}$ denote the set of finite linear combinations of exponential functionals on $(\Omega^H, \mathcal{F}^H, P^H)$. Namely we denote

$$\mathbf{E} = \text{linear span of } \left\{ \exp\left(\int_0^T f(t) dB_t\right), \right.$$

$$\left. f \text{ is a continuously differentiable function on } [0, T] \right\}$$

and

$$\tilde{\mathbf{E}} = \text{linear span of } \left\{ \exp\left(\int_0^T f(t) dB_t^H\right), \right.$$

$$\left. f \text{ is a continuously differentiable function on } [0, T] \right\}.$$

The operators Λ and Γ induce in a natural way transformations defined on $\mathbf{E}$ and $\tilde{\mathbf{E}}$. Set $\mathbf{B} = \Gamma \mathbf{E}$ and $\tilde{\mathbf{B}} = \Lambda \tilde{\mathbf{E}}$.

LEMMA 8.1. *The following diagram*

(8.3)
$$
\begin{array}{ccc}
\mathbf{E} & \xrightarrow{V_{H,T}} & \tilde{\mathbf{E}} \\
\downarrow{\scriptstyle\Gamma} & & \downarrow{\scriptstyle\Lambda} \\
\mathbf{B} & \xrightarrow{V_{H,T}} & \tilde{\mathbf{B}}
\end{array}
$$

commutes.

Proof Denote $\tilde{g} = V_{H,T}^{-1} f$. Then $g = \mathbb{B}_{H,T}\tilde{g}$. Let $F = \exp(\int_0^T h(s)dB_s)$, where $h \in L^2([0,T])$ is a deterministic function. Then

$$
\begin{aligned}
\Gamma F &= \exp\left[\int_0^T h(s)dB_s + \int_0^T h(s)g(s)ds\right] \\
&= \exp\left[\int_0^T h(s)dB_s + \int_0^T h(s)\mathbb{B}_{H,T}\tilde{g}(s)ds\right] \\
&= \exp\left[\int_0^T h(s)dB_s + \int_0^T (\mathbb{B}_{H,T}^* h)(s)\tilde{g}(s)ds\right]
\end{aligned}
$$

Thus

$$
\begin{aligned}
V_{H,T}\Gamma F &= \exp\left[\int_0^T (\mathbb{B}_{H,T}^* h)(s)dB_s^H + \int_0^T (\mathbb{B}_{H,T}^* h)(s)V_{H,T}\tilde{g}(s)ds\right] \\
&= \exp\left[\int_0^T (\mathbb{B}_{H,T}^* h)(s)dB_s^H + \int_0^T (\mathbb{B}_{H,T}^* h)(s)f(s)ds\right].
\end{aligned}
$$

On the other hand,

$$
V_{H,T}F = \exp\left[\int_0^T (\mathbb{B}_{H,T}^* f)(s)dB_s^H\right].
$$

Therefore, we have

$$
\Lambda V_{H,T}F = \exp\left[\int_0^T (\mathbb{B}_{H,T}^* h)(s)dB_s^H + \int_0^T (\mathbb{B}_{H,T}^* h)(s)f(s)ds\right].
$$

This shows that the diagram commutes for exponential functionals. The lemma is proved by a linearity argument. □

LEMMA 8.2. *Let $\Gamma : \Omega \to \Omega$ and $\Lambda : \Omega^H \to \Omega^H$ be measurable and invertible mappings such that the diagram (8.3) commutes. Let $V_{H,T}(F \circ \Gamma) = (V_{H,T}F) \circ \Lambda$ for any $F \in \mathbf{E}$. Then for all $F \in \mathbf{E}$,*

(8.4)
$$
V_{H,T}(F \circ \Gamma^{-1}) = (V_{H,T}F) \circ \Lambda^{-1}.
$$

Proof From $F = F \circ \Gamma^{-1} \circ \Gamma$, it follows that

$$
\begin{aligned}
V_{H,T} \circ F &= V_{H,T} \circ \left[F \circ \Gamma^{-1} \circ \Gamma\right] \\
&= \left[V_{H,T} \circ (F \circ \Gamma^{-1})\right] \circ \Lambda
\end{aligned}
$$

Thus $V_{H,T} \circ (F \circ \Gamma^{-1}) = (V_{H,T} \circ F) \circ \Lambda^{-1}$. □

LEMMA 8.3. *Let the assumptions of Lemma 8.2 be satisfied and let Γ be differentiable in the sense that for any $F \in \mathbb{L}^{1,2}(\Omega^H, \mathcal{F}^H, P^H)$, $F \circ \Gamma$ is in $\mathbb{L}^{1,2}(\Omega^H, \mathcal{F}^H, P^H)$. Then*

$$(8.5) \qquad V_{H,T}(\{D_s(F \circ \Gamma^{-1})\} \circ \Gamma) = \mathbb{I}^*_{H,T}(s)\{D_s^H[(V_{H,T}F) \circ \Lambda^{-1}]\} \circ \Lambda.$$

Proof From Lemma 6.14 and Lemma 8.2 it follows that

$$\begin{aligned}
V_{H,T}(D_s(F \circ \Gamma^{-1}) \circ \Gamma) &= [V_{H,T}D_s(F \circ \Gamma^{-1})] \circ \Lambda \\
&= \mathbb{I}^*_{H,T}(s)D_s^H(V_{H,T}(F \circ \Gamma^{-1})) \circ \Lambda \\
&= \mathbb{I}^*_{H,T}(s)D_s^H[(V_{H,T}F) \circ \Lambda^{-1}] \circ \Lambda.
\end{aligned}$$

This proves the lemma. □

LEMMA 8.4. *Let $\Gamma : \Omega \to \Omega$ and $\Lambda : \Omega^H \to \Omega^H$ be defined by (8.1) and (8.2) respectively. Then for all $F \in \mathbf{E}$,*

$$(8.6) \qquad V_{H,T}(F \circ \Gamma) = (V_{H,T}F) \circ \Lambda.$$

Proof Let $F = \exp\{\int h(s)dB_s\}$. Then

$$\begin{aligned}
F \circ \Gamma &= \exp\left\{\int h(s)dB_s + \int h(s)\mathbb{B}_{H,T}V_{H,T}^{-1}f(s)ds\right\} \\
&= \exp\left\{\int h(s)dB_s + \int (\mathbb{B}^*_{H,T}h)(s)V_{H,T}^{-1}f(s)ds\right\}
\end{aligned}$$

Consequently,

$$\begin{aligned}
V_{H,T}(F \circ \Gamma) &= \exp\left\{\int \mathbb{B}^*_{H,T}h(s)dB_s^H + \int (\mathbb{B}^*_{H,T}h)(s)f(s)ds\right\} \\
&= (V_{H,T}F) \circ \Lambda.
\end{aligned}$$

This proves the lemma for exponential functionals. The lemma follows from a linearity argument. □

The following well-known result is for sBm and can be found for example in **[17]** when the interval is $[0,1]$.

THEOREM 8.5. *Let $g \in \mathbb{L}^{1,2}(\Omega, \mathcal{F}, P)$ satisfy the following conditions.*
(i) There is a positive number $\gamma \in (0,1)$ such that

$$(8.7) \qquad \int_0^T \int_0^T |D_t g(s)|^2 ds dt \qquad \text{is bounded by } \gamma.$$

(ii) There is a positive number $q > 1$ with

$$(8.8) \qquad \mathbb{E}\left[\exp\left\{\frac{q}{2}\int_0^T g^2(s)ds\right\}\right] < \infty.$$

Then Γ defined by (8.2) is invertible and $P \circ \Gamma^{-1}$ is absolutely continuous with respect to P. Moreover, the following identity is true:

$$(8.9) \qquad \frac{dP \circ \Gamma^{-1}}{dP} = \kappa \exp\left\{-\int_0^T g(s)dB_s - \frac{1}{2}\int_0^T g^2(s)ds\right\},$$

with

$$(8.10) \qquad \kappa = \exp\left\{-\int_0^T \int_0^s D_s g(r)\{D_r[g(s, \Gamma_s^{-1})]\} \circ \Gamma_s dr ds\right\},$$

where Γ_s is defined by

$$\Gamma_s : \quad B_{\cdot} \to B_{\cdot} + \int_0^{s\wedge\cdot} g(u)du$$

and Γ_s^{-1} is the inverse of Γ_s.

To obtain the explicit form for Radon-Nikodym derivative for fBm, we define

(8.11) $$h_s(u) := \mathbb{I}_{H,T}\left(I_{[0,s]}\mathbb{B}_{H,T}f\right)(u), \quad 0 \le u \le T$$

and

(8.12) $$\Lambda_s : \quad B_{\cdot}^H \to B_{\cdot}^H + \int_0^{\cdot} h_s(u)du.$$

The main theorem of this section is the following

THEOREM 8.6. *Let f and λ be given as at the beginning of this section and let the following conditions be satisfied*

(i) There is a positive number $\gamma \in (0,1)$ such that

(8.13) $$\int_0^T \int_0^T |\mathbb{I}_{H,T}^*(t)\mathbb{B}_{H,T}(s)D_t^H f(s)|^2 dsdt \qquad \text{is bounded by } \gamma.$$

(ii) There is a positive number $q > 1$ with

(8.14) $$\mathbb{E}\left[\exp\left\{\frac{q}{2}\|\mathbb{B}_{H,T}^*\mathbb{B}_{H,T}f\|_{\Theta_H}^2\right\}\right] < \infty.$$

Then Λ is invertible and $P^H \circ \Lambda^{-1}$ is absolutely continuous with respect to P^H. Moreover, the following identity is true:

(8.15)
$$\frac{dP^H \circ \Lambda^{-1}}{dP^H} = \mathcal{L} = \tilde{\kappa}\exp\left\{-\int_0^T \mathbb{B}_{H,T}^*\mathbb{B}_{H,T}f(s)dB_s^H - \frac{1}{2}\|\mathbb{B}_{H,T}^*\mathbb{B}_{H,T}f\|_{\Theta_H}^2\right\},$$

where

$$\tilde{\kappa} = \exp\left\{-\int_0^T \int_0^s \mathbb{I}_{H,T}^*(s)[D_s^H\mathbb{B}_{H,T}f](r)\mathbb{I}_{H,T}^*(r)\right.$$

(8.16) $$\left.\{D_r^H[\mathbb{B}_{H,T}(f)(s,\Lambda_s)]\} \circ \Lambda_s^{-1}drds\right\},$$

where Λ_s is defined by

$$\Lambda_s : \quad B_{\cdot}^H \to B_{\cdot}^H + \int_0^{\cdot} h_s(u)du$$

and

$$h_s(u) := \mathbb{I}_{H,T}\left(I_{[0,s]}\mathbb{B}_{H,T}f\right)(u), \quad 0 \le u \le T,$$

and Λ_s^{-1} is the inverse of Λ_s.

REMARK 8.7. $\mathbb{I}_{H,T}$, $\mathbb{B}_{H,T}$, $\mathbb{B}_{H,T}^*\mathbb{B}_{H,T}$ have been explicitly computed in Chapter 5.

Proof Let $g = \mathbb{B}_{H,T} V_{H,T}^{-1} f$. By Lemma 6.14 we have

$$
V_{H,T} \left\{ \int_0^T \int_0^T |D_t g(s)|^2 ds dt \right\} = V_{H,T} \int_0^T \int_0^T |D_t \mathbb{B}_{H,T} V_{H,T}^{-1} f|^2 ds dt
$$

$$
= \int_0^T \int_0^T |V_{H,T} D_t \mathbb{B}_{H,T} V_{H,T}^{-1} f|^2 ds dt
$$

$$
= \int_0^T \int_0^T |\mathbb{I}_{H,T}^*(t) D_t^H V_{H,T} \mathbb{B}_{H,T} V_H^{-1} f(s)|^2 ds dt
$$

$$
= \int_0^T \int_0^T |(\mathbb{I}_{H,T}^*(t) D_t^H \mathbb{B}_{H,T} f(s)|^2 ds dt .
$$

Thus (8.13) implies that (8.7) holds. By the property of the probability structure preserving mapping $V_{H,T}$, we obtain

$$
\mathbb{E} \left[\exp \left\{ \frac{q}{2} \int_0^T g^2(s) ds \right\} \right] = \mathbb{E} \left[V_{H,T} \exp \left\{ \frac{q}{2} \int_0^T g^2(s) ds \right\} \right]
$$

$$
= \mathbb{E} \exp \left[\frac{q}{2} \int_0^T (V_{H,T} g)^2(s) ds \right]
$$

$$
= \mathbb{E} \left[\exp \left\{ \frac{q}{2} \int_0^T |\mathbb{B}_{H,T} f(s)|^2 ds \right\} \right]
$$

$$
= \mathbb{E} \left[\exp \left\{ \frac{q}{2} \|\mathbb{B}_{H,T}^* \mathbb{B}_{H,T} f\|_{\Theta_H}^2 \right\} \right] .
$$

Hence (8.14) implies that (8.8) is true. Therefore under the assumptions of Theorem 8.5, $\dfrac{dP \circ \Gamma^{-1}}{dP}$ exists and is given by (8.9)-(8.10). From Theorem 2.11 of [**48**] we conclude that $\frac{dP^H \circ \Lambda^{-1}}{dP^H}$ exists and is given by

$$
\frac{dP^H \circ \Lambda^{-1}}{dP^H} = V_{H,T} \circ \left(\frac{dP \circ \Gamma^{-1}}{dP} \right) .
$$

Now we want to compute $V_{H,T} \circ \left(\frac{dP \circ \Gamma^{-1}}{dP} \right)$. By proposition 6.13 we have

$$
V_{H,T} \exp \left\{ - \int_0^T g(s) dB_s - \frac{1}{2} \int_0^T g^2(s) ds \right\}
$$

$$
= \exp \left\{ - \int_0^T \mathbb{B}_{H,T}^* V_{H,T} g(s) dB_s^H - \frac{1}{2} \int_0^T V_{H,T} g^2(s) ds \right\}
$$

$$
= \exp \left\{ - \int_0^T \mathbb{B}_{H,T}^* \mathbb{B}_{H,T} f(s) dB_s^H - \frac{1}{2} \int_0^T |\mathbb{B}_{H,T} f|^2(s) ds \right\}
$$

$$
= \exp \left\{ - \int_0^T \mathbb{B}_{H,T}^* \mathbb{B}_{H,T} f(s) dB_s^H - \frac{1}{2} \|\mathbb{B}_{H,T}^* \mathbb{B}_{H,T} f\|_{\Theta_H}^2 \right\}
$$

Next we need to evaluate $\tilde{\kappa} = V_{H,T}\kappa$. Let Γ_s be defined as in Theorem 8.5 and let Λ_s be defined as in Theorem 8.6. First we prove that the following diagram

(8.17)
$$
\begin{array}{ccc}
\mathbf{E} & \overset{V_{H,T}}{\longrightarrow} & \tilde{\mathbf{E}} \\
\downarrow{\scriptstyle \Gamma_s} & & \downarrow{\scriptstyle \Lambda_s} \\
\mathbf{B} & \overset{V_{H,T}}{\longrightarrow} & \tilde{\mathbf{B}}
\end{array}
$$

commutes. In fact, by Lemma 6.14 we have

$$
\begin{aligned}
V_{H,T}(D_s g(r)) &= \mathbb{I}^*_{H,T}(s) D_s^H \left[V_{H,T} g(r) \right] \\
&= \mathbb{I}^*_{H,T}(s) D_s^H \mathbb{B}_{H,T}(f)(r) .
\end{aligned}
$$

If $F = \exp\left(\int_0^T h(s)dB_s \right)$, where h is continuously differentiable on $[0,T]$, then

$$
F \circ \Gamma_s = \exp\left(\int_0^T h(u)dB_u + \int_0^s h(u)g(u)du \right)
$$

and consequently

$$
V\left(F \circ \Gamma_s \right) = \exp\left(\int_0^T \mathbb{B}^*_{H,T} h(u) dB_u^H + \int_0^s h(u) V_{H,T} g(u) du \right) .
$$

On the other hand we have

$$
V_{H,T} F = \exp\left(\int_0^T \mathbb{B}^*_{H,T} h(u) dB_u^H \right) .
$$

Thus

$$
\begin{aligned}
(V_{H,T}F) \circ \Lambda_s &= \exp\left(\int_0^T \mathbb{B}^*_{H,T} h(u) dB_u^H + \int_0^T \mathbb{B}^*_{H,T} h(u) \mathbb{I}_{H,T}\left(I_{[0,s]} \mathbb{B}_{H,T} f \right)(u) du \right) \\
&= \exp\left(\int_0^T \mathbb{B}^*_{H,T} h(u) dB_u^H + \int_0^s h(u) \mathbb{B}_{H,T} f(u) du \right) \\
&= \exp\left(\int_0^T \mathbb{B}^*_{H,T} h(u) dB_u^H + \int_0^s h(u) V_{H,T} g(u) du \right) .
\end{aligned}
$$

This shows that

$$
V_{H,T}\left(F \circ \Gamma_s \right) = (V_{H,T}F) \circ \Lambda_s .
$$

Therefore from Lemma 8.3 it follows that

$$
\begin{aligned}
V_{H,T}\left(D_r(g(s,\Gamma_s^{-1})) \circ \Gamma_s \right) &= \mathbb{I}^*_{H,T}(r) \left\{ D_r^H \left[(V_{H,T}g) \circ \Lambda_s^{-1} \right] \right\} \circ \Lambda_s \\
&= \mathbb{I}^*_{H,T}(r) \left\{ \left[D_r^H (\mathbb{B}_{H,T}f)(s,\Lambda_s^{-1}) \right] \right\} \circ \Lambda_s .
\end{aligned}
$$

Thus

$$
V_{H,T}\kappa = \exp\left\{ -\int_0^T \int_0^s \mathbb{I}^*_{H,T}(s)[D_s^H \mathbb{B}_{H,T}f](r)\mathbb{I}^*_{H,T}(r) \right.
$$

(8.18)
$$
\left. \{D_r^H[\mathbb{B}_{H,T}f(s,\Lambda_s^{-1})]\} \circ \Lambda_s dr ds \right\} .
$$

This proves the theorem. $\qquad\square$

If f is deterministic, then $D_s f = 0$. Thus $\tilde{\kappa} \equiv 1$. The following corollary is obvious.

COROLLARY 8.8. If f is deterministic and $\mathbb{B}_{H,T}^* \mathbb{B}_{H,T} f \in \Theta_H$, then

$$(8.19) \qquad \frac{dP^H \circ \Lambda^{-1}}{dP^H} = \exp\left\{ -\int_0^T \mathbb{B}_{H,T}^* \mathbb{B}_{H,T} f(s) dB_s^H - \frac{1}{2} \|\mathbb{B}_{H,T}^* \mathbb{B}_{H,T} f\|_{\Theta_H}^2 \right\}.$$

We give examples to compute $\mathbb{B}_{H,T}^* \mathbb{B}_{H,T} f$.

Let the translation be given by $f(s) = as^\beta$, where a and β are constants.

(i) If $0 < H < 1/2$ and $\beta > H - \frac{3}{2}$, then by (5.38) and (5.39), we have

$$
\begin{aligned}
\mathbb{B}_{H,T}^* \mathbb{B}_{H,T} f(t) &= \frac{H^2(1-2H)^2 \kappa_1^2}{\kappa_H^2} t^{\frac{1}{2}-H} \\
&\quad \int_t^T r^{2H-1}(r-t)^{-\frac{1}{2}-H} \int_0^r u^{\frac{1}{2}+\beta-H}(r-u)^{-\frac{1}{2}-H} du \\
&= \frac{H^2(1-2H)^2 \kappa_1^2 B(\frac{3}{2}+\beta-H, \frac{1}{2}-H)}{\kappa_H^2} \\
(8.20) &\quad t^{\frac{1}{2}-H} \int_t^T r^\beta (r-t)^{-\frac{1}{2}-H} dr.
\end{aligned}
$$

In particular if $0 < H < 1/2$ and $f = a$, i.e. $\beta = 0$, then

$$
\begin{aligned}
\mathbb{B}_{H,T}^* \mathbb{B}_{H,T} f(t) &= a \frac{\kappa_1^2 B(\frac{3}{2}-H, \frac{1}{2}-H) H^2(1-H)^2}{\kappa_H^2(\frac{1}{2}-H)} t^{\frac{1}{2}-H}(T-t)^{\frac{1}{2}-H} \\
(8.21) &= \frac{a}{2H\Gamma(\frac{1}{2}+H)\Gamma(\frac{3}{2}-H)} t^{\frac{1}{2}-H}(T-t)^{\frac{1}{2}-H}.
\end{aligned}
$$

(ii) $1/2 < H < 1$ and $\beta > H - \frac{3}{2}$, then by (5.36), we have

$$
\begin{aligned}
\mathbb{B}_{H,T}^* \mathbb{B}_{H,T} f(t) &= -\kappa_H t^{1/2-H} \frac{d}{dt} \int_t^T dw\, w^{2H-1}(w-t)^{1/2-H} \\
&\quad \frac{d}{dw} \int_0^w z^{1/2-H+\beta}(w-z)^{1/2-H} dz \\
&= -\kappa_H B(\frac{3}{2}-H+\beta, \frac{3}{2}-H)(2-H+\beta) t^{\frac{1}{2}-H} \\
(8.22) &\quad \frac{d}{dt} \int_t^T w^\beta (w-t)^{1/2-H} dw.
\end{aligned}
$$

In particular if $0 < H < 1/2$ and $f = a$, i.e. $\beta = 0$, then

$$
\begin{aligned}
\mathbb{B}_{H,T}^* \mathbb{B}_{H,T} f(t) &= -\frac{a\kappa_H B(\frac{3}{2}-H, \frac{3}{2}-H)(2-H)}{\frac{3}{2}-H} t^{\frac{1}{2}-H}(T-t)^{3/2-H} \\
(8.23) &= -\frac{\Gamma(\frac{3}{2}-H)\Gamma(H+\frac{1}{2})}{H(3-2H)} t^{\frac{1}{2}-H}(T-t)^{3/2-H}.
\end{aligned}
$$

COROLLARY 8.9. (a) Let $0 < H < 1/2$ and let $f(s) = as^\beta$, where a and β are constants with $\beta > H - \frac{3}{2}$. Consider the translation

$$\Lambda: B_\cdot^H \to B_\cdot^H + \int_0^\cdot f(s) ds.$$

Then $P^H \circ \Lambda^{-1}$ is absolutely continuous with respect to P^H and

$$(8.24) \qquad \frac{dP^H \circ \Lambda^{-1}}{dP^H} = \exp\left(\int_0^T h(s)dB_s^H - \frac{1}{2}\|h\|_{\Theta_H}^2\right),$$

where

$$(8.25) \quad h(s) = \frac{H^2(1-2H)^2\kappa_1^2 B(\frac{3}{2}+\beta-H,\frac{1}{2}-H)}{\kappa_H^2} t^{\frac{1}{2}-H} \int_t^T r^\beta (r-t)^{-\frac{1}{2}-H} dr.$$

In particular if $f = a$, then

$$(8.26) \qquad h(s) = \frac{a}{2H\Gamma(\frac{1}{2}+H)\Gamma(\frac{3}{2}-H)} t^{\frac{1}{2}-H}(T-t)^{\frac{1}{2}-H}.$$

(b) Let $\frac{1}{2} < H < 1$, $\beta > H - \frac{3}{2}$ and put $f(s) = as^\beta$. Consider the translation

$$\Lambda: B_\cdot^H \to B_\cdot^H + \int_0^\cdot f(s)ds.$$

Then $P^H \circ \Lambda^{-1}$ is absolutely continuous with respect to P^H and

$$(8.27) \qquad \frac{dP^H \circ \Lambda^{-1}}{dP^H} = \exp\left(\int_0^T h(s)dB_s^H - \frac{1}{2}\|h\|_{\Theta_H}^2\right),$$

where

$$(8.28) \quad h(s) = -\kappa_H B(\frac{3}{2}-H+\beta, \frac{3}{2}-H)(2-H+\beta)t^{\frac{1}{2}-H} \frac{d}{dt} \int_t^T w^\beta (w-t)^{1/2-H} dw.$$

In particular if $f = a$, then

$$(8.29) \qquad h(s) = -\frac{\Gamma(\frac{3}{2}-H)\Gamma(H+\frac{1}{2})}{H(3-2H)} t^{\frac{1}{2}-H}(T-t)^{3/2-H}.$$

(c) If $T = \infty$, then a similar results has been proved in [**48**].

REMARK 8.10. (a) The Girsanov theorem given in [**33**] (Theorem 4.9) imposes the condition that $\mathbb{E}_H[\Lambda_1^u] = 1$ for some functional Λ_1^u. This conditional may be implied by the famous Novikov condition. However, first, it is assumed the adaptedness of the translation process. Secondly, the stochastic integral $\int_0^t u_s \delta_H B_s^H$ introduced in [**33**] is different to the definition introduced in this paper or that of [**35**] (see Chapter 6.5). Third, their Girsanov theorem is different type than the one we prove in this paper.

(b) The case $f(s) = a$ in the above corollary was obtained in [**90**]. Hence this corollary extends the work of [**90**].

CHAPTER 9

Conditional Expectation

Let F be a (nonlinear) functional of fBm $(B_t^H, t \geq 0)$ and let $\mathcal{F}_T = \sigma(B_t^H, 0 \leq t \leq T)$ be the σ-algebra generated by $(B_t^H, 0 \leq t \leq T)$. If $F = \int_0^\infty f(s)dB_s^H$, $f \in \Theta_{H,\infty}$, then it is known ([47]) that

(9.1)
$$\mathbb{E}\left[F|\mathcal{F}_T\right] = \int_0^T h(t)dB_t^H ,$$

where $h \in \Theta_{H,T}$ satisfies the following fractional differential (or integral) equation

(9.2)
$$\begin{cases} (-\Delta)^\beta h_T(s) = (-\Delta)^\beta f(s) & \forall\, s \in [0,T] \\ h_T(s) = 0 & \forall\, s \notin [0,T], \end{cases}$$

where $\Delta = \dfrac{d^2}{dx^2}$ is the one dimensional Laplacian and $\beta = \frac{1}{2} - H$. If $H > 1/2$, then this equation is a fractional integral equation, called the *Carleman equation* (c.f. [76], [106] and the references therein). It can be solved explicitly by integral kernel. If $H < 1/2$, then it is a fractional differential equation and an explicit solution is given in [47].

In this section, first we give an alternative way to find $\mathbb{E}\left[F\Big|\mathcal{F}_T\right]$. It is easy to see that there is an $h \in \Theta_{H,T}$ such that

$$\mathbb{E}\left[F|\mathcal{F}_T\right] = \int_0^T h(t)dB_t^H .$$

To determine h in the above identity we use the fact that for any $g \in \mathbf{S}$ the following holds

$$\mathbb{E}\left[\int_0^\infty f(s)dB_s^H \int_0^T g(t)dB_t^H\right] = \mathbb{E}\left[\int_0^T h(s)dB_s^H \int_0^T g(t)dB_t^H\right] .$$

By the definition of stochastic integral the above equality can be written as

$$\mathbb{E}\left[\int_0^\infty \mathbb{I}_{H,\infty}^* f(s)dB_s \int_0^T \mathbb{I}_{H,T}^* g(s)dB_s\right] = \mathbb{E}\left[\int_0^T \mathbb{I}_{H,T}^* h(s)dB_s \int_0^T \mathbb{I}_{H,T}^* g(s)dB_s\right] .$$

Thus the above equation reduces to

$$\int_0^T \mathbb{I}_{H,\infty}^* f(t)\mathbb{I}_{H,T}^* g(t)dt = \int_0^T \mathbb{I}_{H,T}^* h(t)\mathbb{I}_{H,T}^* g(t)dt$$

or

$$\int_0^T \mathbb{I}_{H,T}\mathbb{I}_{H,T}^* h(t)g(t)dt = \int_0^T \mathbb{I}_{H,T}\mathbb{I}_{H,\infty}^* f(t)g(t)dt$$

Since $g \in \mathbf{S}$ is arbitrary, we have

$$\mathbb{I}_{H,T}\mathbb{I}^*_{H,T}h(t) = \mathbb{I}_{H,T}\mathbb{I}^*_{H,\infty}f(t), \quad \forall\, 0 \le t \le T.$$

Therefore it follows that

$$
\begin{aligned}
h(t) &= \mathbb{P}_{H,T}f(t) := \left(\mathbb{I}_{H,T}\mathbb{I}^*_{H,T}\right)^{-1}\mathbb{I}_{H,T}\mathbb{I}^*_{H,\infty}f(t) \\
&= \mathbb{B}^*_{H,T}\mathbb{B}_{H,T}\mathbb{I}_{H,T}\mathbb{I}^*_{H,\infty}f(t) \\
&= \mathbb{B}^*_{H,T}\mathbb{I}^*_{H,\infty}f(t).
\end{aligned}
$$

(9.3)

Let f be continuously differentiable we give an expression to compute $\mathbb{B}^*_{H,T}\mathbb{I}^*_{H,T}f$. First let $0 < H < 1/2$. Then

$$
\begin{aligned}
&\mathbb{B}^*_{H,T}\mathbb{I}^*_{H,\infty}f(t) \\
={} & -H(1-2H)\kappa_1 t^{\frac{1}{2}-H}\int_t^T (s-t)^{-\frac{1}{2}-H}\left[\frac{d}{ds}\int_s^\infty r^{H-\frac{1}{2}}(r-s)^{H-\frac{1}{2}}f(r)dr\right]ds \\
={} & -H(1-2H)\kappa_1 t^{\frac{1}{2}-H}\int_t^T (s-t)^{-\frac{1}{2}-H}\int_s^\infty (r-s)^{H-\frac{1}{2}}\frac{d}{dr}\left(r^{H-\frac{1}{2}}f(r)\right)drds \\
={} & -H(1-2H)\kappa_1 t^{\frac{1}{2}-H}\int_t^T\int_t^r (s-t)^{-\frac{1}{2}-H}(r-s)^{H-\frac{1}{2}}ds\frac{d}{dr}\left(r^{H-\frac{1}{2}}f(r)\right)dr \\
& -H(1-2H)\kappa_1 t^{\frac{1}{2}-H}\int_T^\infty\int_t^T (s-t)^{-\frac{1}{2}-H}(r-s)^{H-\frac{1}{2}}ds\frac{d}{dr}\left(r^{H-\frac{1}{2}}f(r)\right)dr \\
={} & -H(1-2H)B(\tfrac{1}{2}-H,\tfrac{1}{2}+H)\kappa_1 t^{\frac{1}{2}-H}\int_t^T \frac{d}{dr}\left(r^{H-\frac{1}{2}}f(r)\right)dr \\
& -H(1-2H)\kappa_1 t^{\frac{1}{2}-H}\int_T^\infty \theta_{H,T,r}\frac{d}{dr}\left(r^{H-\frac{1}{2}}f(r)\right)dr \\
={} & f(t) - t^{\frac{1}{2}-H}T^{H-\frac{1}{2}}f(T) \\
& -t^{\frac{1}{2}-H}\int_T^\infty \theta_{H,T,r}\frac{d}{dr}\left(r^{H-\frac{1}{2}}f(r)\right)dr
\end{aligned}
$$

where

$$
\begin{aligned}
\theta_{H,T,r} &= H(1-2H)\kappa_1\int_t^T (s-t)^{-\frac{1}{2}-H}(r-s)^{H-\frac{1}{2}}ds \\
&= H(1-2H)\kappa_1\left(\frac{T-t}{r-t}\right)^{\frac{1}{2}-H}\int_0^{T-t} x^{-\frac{1}{2}-H}(1-\frac{T-t}{r-t}x)^{H-\frac{1}{2}}dx.
\end{aligned}
$$

The above last integral is the integral representation of a hypergeometric function of $\frac{T-t}{r-t}$. Using the conventional notation, we have

$$\theta_{H,T,r} = F(\tfrac{1}{2}-H,\tfrac{1}{2}-H,\tfrac{3}{2}-H,\frac{T-t}{r-t}).$$

Now let $1/2 < H < 1$. Then

$$\mathbb{B}^*_{H,T}\mathbf{I}^*_{H,\infty}f(t)$$

$$= -\frac{1}{\Gamma(H+\frac{1}{2})\Gamma(\frac{3}{2}-H)\kappa_H}t^{\frac{1}{2}H}\frac{d}{dt}\int_t^T u^{II-\frac{1}{2}}(u-t)^{\frac{1}{2}-H}$$

$$(-\kappa_H)u^{\frac{1}{2}-H}\frac{d}{du}\int_u^\infty v^{H-\frac{1}{2}}(v-u)^{H-\frac{1}{2}}f(v)dvdu$$

$$= \frac{1}{\Gamma(H+\frac{1}{2})\Gamma(\frac{3}{2}-H)}t^{\frac{1}{2}-H}\frac{d}{dt}\int_t^\infty \left[\int_t^{v\wedge T}(u-t)^{\frac{1}{2}-H}\right.$$

$$\left. (v-u)^{H-\frac{1}{2}}\frac{d}{dv}\left(v^{H-\frac{1}{2}}f(v)\right)du\right]dv$$

$$= t^{\frac{1}{2}-H}\frac{d}{dt}\int_t^T (v-t)\frac{d}{dv}\left(v^{H-\frac{1}{2}}f(v)\right)dv$$

$$+\frac{1}{\Gamma(H+\frac{1}{2})\Gamma(\frac{3}{2}-H)}t^{\frac{1}{2}-H}\frac{d}{dt}\int_T^\infty \int_t^T (u-t)^{\frac{1}{2}-H}(v-u)^{H-\frac{1}{2}}du$$

$$\frac{d}{dv}\left(v^{H-\frac{1}{2}}f(v)\right)dv$$

$$= f(t) - t^{\frac{1}{2}-H}T^{H-\frac{1}{2}}f(T)$$

$$-t^{\frac{1}{2}-H}\int_T^\infty \theta_{H,T,t}(v)\frac{d}{dv}\left(v^{H-\frac{1}{2}}f(v)\right)dv,$$

where

$$(9.4) \quad \theta_{H,T,t}(v) = (T-t)^{\frac{1}{2}-H}(v-T)^{H-\frac{1}{2}} + \left(H-\frac{1}{2}\right)\int_0^{T-t} r^{\frac{1}{2}-H}(v-t-r)^{H-\frac{3}{2}}dr.$$

THEOREM 9.1. *Let $F = \int_0^\infty f(t)dB_t^H$, where $f \in \Theta_{H,\infty}$ is deterministic. Then*

$$(9.5) \qquad\qquad \mathbb{E}\left[F|\mathcal{F}_T\right] = \int_0^T h(t)dB_t^H,$$

where

$$(9.6) \quad h(t) = f(t) - t^{\frac{1}{2}-H}T^{H-\frac{1}{2}}f(T) - t^{\frac{1}{2}-H}\int_T^\infty \theta_{H,T,t}(v)\frac{d}{dv}\left(v^{H-\frac{1}{2}}f(v)\right)dv,$$

where
$$(9.7)$$

$$\theta_{H,T,t}(v) = \begin{cases} (T-t)^{\frac{1}{2}-H}(v-T)^{H-\frac{1}{2}} + (H-\frac{1}{2})\int_0^{T-t} r^{\frac{1}{2}-H}(v-t-r)^{H-\frac{3}{2}}dr \\ \qquad\qquad if\ 1/2 < H < 1 \\ H(1-2H)\kappa_1\left(\frac{T-t}{r-t}\right)^{\frac{1}{2}-H}\int_0^{T-t} x^{-\frac{1}{2}-H}(1-\frac{T-t}{r-t}x)^{H-\frac{1}{2}}dx \\ \qquad\qquad if\ 0 < H < 1/2 \end{cases}$$

This theorem can be used to compute the conditional expectation of linear functional of fBm. To compute the conditional expectation of general (nonlinear) functional of fBm we shall first compute the conditional expectation of exponential functionals of fBm. Let $f \in \mathbf{S}$. Now we compute $\mathbb{E}\left[\exp\left\{\int_0^\infty f(t)dB_t^H - \frac{1}{2}\|f\|^2_{\Theta_{H,T}}\right\}|\mathcal{F}_T\right]$. The following lemma gives the formula to compute it.

LEMMA 9.2. *Let* $f \in \Theta_{H,\infty}$ *be deterministic. Then* h *defined by (9.3) (or (9.6))* *is in* $\Theta_{H,T}$ *and*

(9.8)
$$\mathbb{E}\left[\exp\left\{\int_0^\infty f(t)dB_t^H - \frac{1}{2}\|f\|^2_{\Theta_{H,\infty}}\right\}\Big|\mathcal{F}_T\right] = \exp\left\{\int_0^T h(t)dB_t^H - \frac{1}{2}\|h\|^2_{\Theta_{H,T}}\right\}.$$

Proof Since $f \in \Theta_{H,\infty}$, $\mathbb{I}^*_{H,\infty}f$ exists and is in $L^2([0,\infty))$. The restriction of $\mathbb{I}^*_{H,\infty}f$ to the interval $[0,T]$ is in $L^2([0,T])$. Therefore $\mathbb{B}^*_{H,T}\mathbb{I}^*_{H,\infty}f$ is in $\Theta_{H,T}$. Now it suffices to show that for any continuously differentiable function g on $[0,T]$, we have

(9.9)
$$\mathbb{E}\left[\exp\left\{\int_0^\infty f(s)dB_s^H - \frac{1}{2}\|f\|^2_{\Theta_H}\right\}e^{\int_0^T g(s)dB_s^H}\right]$$
$$= \mathbb{E}\left[\exp\left\{\int_0^T h(s)dB_s^H - \frac{1}{2}\|h\|^2_{\Theta_H}\right\}e^{\int_0^T g(s)dB_s^H}\right].$$

The left hand side of (9.9) is

$$\exp\left\{\frac{1}{2}\|g\|^2_{\Theta_{H,T}} + \langle f, gI_{[0,T]}\rangle_{\Theta_{H,\infty}}\right\}$$

The right hand side of (9.9) is

$$\exp\left\{\frac{1}{2}\|g\|^2_{\Theta_{H,T}} + \langle h, g\rangle_{\Theta_{H,T}}\right\}.$$

Equation (9.9) is equivalent to

$$\langle f, I_{[0,T]}g\rangle_{\Theta_{H,T}} = \langle h, g\rangle_{\Theta_{H,T}}$$

which is true by the definition of h. This proves the lemma. $\qquad\square$

From this lemma, we see that for any $f \in \Theta_{H,\infty}$

(9.10)
$$\mathbb{E}\left[\exp\left\{x\int_0^\infty f(t)dB_t^H - \frac{x^2}{2}\|f\|^2_{\Theta_{H,T}}\right\}\Big|\mathcal{F}_T\right] = \exp\left\{x\int_0^T h(t)dB_t^H - \frac{x^2}{2}\|h\|^2_{\Theta_{H,T}}\right\},$$

where $h = \mathbb{B}^*_{H,T}\mathbb{I}^*_{H,\infty}f$. Expanding it as a function of x, we have

$$\sum_{n=0}^\infty \frac{x^n}{n!}\mathbb{E}\left[I_n^{H,\infty}(f^{\otimes n})\big|\mathcal{F}_T\right] = \sum_{n=0}^\infty \frac{x^n}{n!}I_n^{H,T}(h^{\otimes n}).$$

By comparing the coefficients of x^n, $n = 0, 1, 2, \cdots$

$$\mathbb{E}\left[I_n^{H,\infty}(f^{\otimes n})\big|\mathcal{F}_T\right] = I_n^{H,T}((\mathbb{P}_{H,T}f)^{\otimes n}),$$

where

(9.11) $$\mathbb{P}_{H,T} = \mathbb{B}^*_{H,T}\mathbb{I}^*_{H,\infty}.$$

Since conditional expectation operation is linear and by polarization technique, we have

THEOREM 9.3. *Let* $f_n \in \Theta_{H,\infty}^{\otimes n}$. *Then*

(9.12) $$\mathbb{E}\left[I_n^{H,\infty}(f_n)\big|\mathcal{F}_T\right] = I_n^{H,T}\left(\mathbb{P}_{H,T}^{\otimes n}f_n\right).$$

Write it in multiple stochastic integral form we have

$$(9.13) \quad \mathbb{E}\left[\int_{0<t_1,t_2,\cdots,t_n<\infty} f_n(t_1,\cdots,t_n)dB_{t_1}^H\cdots dB_{t_n}^H \Big| \mathcal{F}_T\right]$$
$$= \int_{0<t_1,t_2,\cdots,t_n<T} h_n(t_1,\cdots,t_n)dB_{t_1}^H\cdots dB_{t_n}^H,$$

where

$$(9.14) \quad h_n(t_1,\cdots,t_n) = \mathbb{P}_{H,T}^{\otimes n} f_n(t_1,\cdots,t_n).$$

Let $\{\eta_1,\eta_2,\cdots,\eta_n,\cdots\} \subset \mathbf{S}([0,\infty))$ be an orthonormal basis of $\Theta_H = \Theta_H([0,\infty))$. Let $f_n \in \Theta_H^{\otimes n}$. If the limit of

$$\sum_{n_1,\cdots,n_k=1}^m \langle f_n, \eta_{n_1}\otimes\eta_{n_1}\otimes\eta_{n_2}\otimes\eta_{n_2}\otimes\cdots\otimes\eta_{n_k}\otimes\eta_{n_k}\rangle_{\Theta_H^{\otimes 2k}}$$

exists as an element of $\Theta_{H,T}^{\otimes(n-2k)}$ as $m\to\infty$, then we say that the k-trace of f_n exists and we denote the above limit by $\mathrm{Tr}^k f_n$.

We also need the following concept of (k,l)-trace. Let $\{\xi_1,\cdots,\xi_n,\cdots\} \subset \mathbf{S}([0,T])$ be an orthonormal basis of $\Theta_{H,T}$. We consider $\{\xi_1,\cdots,\xi_n,\cdots\}$ as a subset of $\Theta_{H,\infty}$. Denote $\zeta_n = \mathrm{I}\!\Gamma_{H,\infty}\mathbb{B}_{H,T}\xi_n$. If the following limit

$$\lim_{m\to\infty} \sum_{n_1,\cdots,n_k,\cdots,n_{k+l}=1}^m \langle f_n, \eta_{n_1}\otimes\eta_{n_1}\otimes\eta_{n_2}\otimes\eta_{n_2}\otimes\cdots\otimes\eta_{n_k}\otimes\eta_{n_k}$$
$$\otimes\zeta_{n_{k+1}}\otimes\zeta_{n_{k+1}}\otimes\cdots\otimes\zeta_{n_{k+l}}\otimes\zeta_{n_{k+l}}\rangle_{\Theta_{H,T}^{\otimes 2k}}$$

exists as an element of $\Theta_{H,T}^{\otimes(n-2k-2l)}$, then we say that the (k,l)-trace of f_n exists and we denote the above limit by $\mathrm{Tr}^{k,l} f_n$.

DEFINITION 9.4. If $f_n \in \Theta_{H,T}^{\otimes n}$ and if the k traces of f_n exist for all $k \leq n/2$, then we say that the multiple (Stratonovich type) integral of f_n is well-defined and we define this multiple integral as

$$(9.15) \quad S_n^{H,T}(f_n) = \sum_{k\leq n/2} \frac{n!}{2^k k!(n-2k)!} I_{n-2k}^{H,T}(\mathrm{Tr}^k f_n).$$

REMARK 9.5. (a) The multiple stochastic integrals of Stratonovich type (with respect to sBm) have been studied in [**52**], [**53**], and [**54**] and its relation with multiple Itô-Wiener integrals (with respect to sBm) is related to "trace". Later on this relation has been extensively studied in [**16**], [**29**], [**65**], [**66**], [**102**], and [**103**] (see also the references therein). Multiple Stratonovich integrals and multiple Itô integrals with respect to fBm have been also studied briefly in [**27**], [**28**], and [**35**].

(b) We define the multiple stochastic integral of Stratonovich type using the Hu-Meyer like formula ([**52**]-[**54**]).

It is easy to check that if $f \in \Theta_{H,T}$, then $f^{\otimes n}$ has all traces. Moreover, it is straightforward to verify that

$$(9.16) \quad S_n^{H,T}(f^{\otimes n}) = \left(\int_0^T f(t)dB_t^H\right)^n$$

and then

$$(9.17) \qquad e^{\int_0^T f(t)dB_t^H} = \sum_{n=0}^{\infty} \frac{1}{n!} S_n^{H,T}(f^{\otimes n}).$$

Now we are going to compute the $\mathbb{E}\left[S_n^{H,T}(f_n)|\mathcal{F}_t\right]$ for all f_n which has all traces. Recall the definition of $\mathbb{P}_{H,T}f(t)$ by (9.11). Equation (9.8) also yields

(9.18)
$$\mathbb{E}\left[\exp\left\{x\int_0^{\infty} f(t)dB_t^H\right\}|\mathcal{F}_T\right] = \exp\left\{x\int_0^T h(t)dB_t^H + \frac{x^2}{2}\left[\|f\|_{\Theta_{H,\infty}}^2 - \|h\|_{\Theta_{H,T}}^2\right]\right\},$$

where $h = \mathbb{P}_{H,T}f$. Expanding it in terms of a power series of x^n, $n = 0, 1, 2, \cdots$, we have

$$\sum_{n=0}^{\infty} \frac{x^n}{n!} \mathbb{E}\left[S_n^{H,\infty}(f^{\otimes n})|\mathcal{F}_T\right]$$

$$= \sum_{n=0}^{\infty} x^n \sum_{k \leq n/2} \frac{1}{2^k k!(n-2k)!} S_{n-2k}^{H,T}(h^{\otimes(n-2k)}) \left(\|f\|_{\Theta_{H,\infty}}^2 - \|h\|_{\Theta_{H,T}}^2\right)^k$$

$$= \sum_{n=0}^{\infty} x^n \sum_{k \leq n/2} \frac{1}{2^k k!(n-2k)!} S_{n-2k}^{H,T}(h^{\otimes(n-2k)}) \sum_{l=0}^{k} (-1)^l \binom{k}{l} \langle f, f \rangle_{\Theta_{H,\infty}}^{k-l} \langle h, h \rangle_{\Theta_{H,T}}^l.$$

Comparing the coefficients of x^n, we have

$$\mathbb{E}\left[S_n^{H,\infty}(f^{\otimes n})|\mathcal{F}_T\right]$$

$$= \sum_{k \leq n/2} \sum_{l=0}^{k} \frac{(-1)^l n!}{2^k l!(k-l)!(n-2k)!} \langle f, f \rangle_{\Theta_{H,\infty}}^{k-l} \langle h, h \rangle_{\Theta_{H,T}}^l S_{n-2k}^{H,T}(h^{\otimes(n-2k)}).$$

It is easy to obvious that

$$\langle f, f \rangle_{\Theta_{H,\infty}} = \sum_{k=1}^{\infty} \langle f \otimes f, \eta_n \otimes \eta_n \rangle_{\Theta_{H,\infty}^{\otimes 2}}$$

and

$$\langle h, h \rangle_{\Theta_{H,T}} = \sum_{k=1}^{\infty} \langle h \otimes h, \xi_n \otimes \xi_n \rangle_{\Theta_{H,T}^{\otimes 2}}$$

$$= \sum_{k=1}^{\infty} \langle \mathbb{B}_{H,T}^* \mathbb{I}_{H,\infty}^* f \otimes \mathbb{B}_{H,T}^* \mathbb{I}_{H,\infty}^* f, \xi_n \otimes \xi_n \rangle_{\Theta_{H,T}^{\otimes 2}}$$

$$= \sum_{k=1}^{\infty} \langle f \otimes f, \mathbb{I}_{H,\infty} \mathbb{B}_{H,T} \xi_n \otimes \mathbb{I}_{H,\infty} \mathbb{B}_{H,T} \xi_n \rangle_{\Theta_{H,T}^{\otimes 2}}$$

$$= \sum_{k=1}^{\infty} \langle f \otimes f, \zeta_n \otimes \zeta_n \rangle_{\Theta_{H,T}^{\otimes 2}}.$$

Thus

$$\mathbb{E}\left[S_n^{H,\infty}(f^{\otimes n})|\mathcal{F}_T\right]$$

$$= \sum_{k \leq n/2} \sum_{l=0}^{k} \frac{(-1)^l n!}{2^k l!(k-l)!(n-2k)!} S_{n-2k}^{H,T}(\mathbb{P}_{H,T}^{\otimes(n-2k)} \mathrm{Tr}^{k-l,l} f^{\otimes n}).$$

THEOREM 9.6. *Let* $f_n \in \Theta_{H,\infty}^{\otimes n}$ *and let the* $(k-l,l)$*-traces of* f_n *exists for all* $0 \le l \le k \le n/2$. *Then* $S_n(f_n)$ *is well-defined and is in* $L^2(\Omega, \mathcal{F}, P^H)$. *Furthermore, we have*
(9.19)
$$\mathbb{E}\left[S_n(f_n)|\mathcal{F}_T\right] = \sum_{k \le n/2}\sum_{l=0}^{k}\frac{(-1)^l}{2^k l!(k-l)!(n-2k)!}S_{n-2k}\left(\mathbb{P}_{H,T}^{\otimes(n-2k)}\ Tr^{k-l,l}f_n\right).$$

EXAMPLE 9.7. Let $f \in \Theta_{H,\infty}$. Then
(9.20)
$$\mathbb{E}\left[\left(\int_0^\infty f(t)dB_t^H\right)^n\Big|\mathcal{F}_T\right]$$
$$= \sum_{k \le n/2}\sum_{l=0}^{k}\frac{(-1)^l n!}{2^k l!(k-l)!(n-2k)!}\|f\|_{\Theta_{H,\infty}}^{k-l}\|h\|_{\Theta_{H,T}}^{l}\left(\int_0^T h(t)dB_t^H\right)^{n-2k},$$
where $h = \mathbb{P}_{H,T}f$.

Now we study the conditional expectation of stochastic integral.

THEOREM 9.8. *Let* f *be algebraically integrable. Then*
(9.21)
$$\mathbb{E}\left[\int_0^T f(s)dB_s^H\Big|\mathcal{F}_\tau\right] = \int_0^\tau \mathbb{P}_{H,\tau}(s)\mathbb{E}\left[f(s)\Big|\mathcal{F}_\tau\right]dB_s^H,$$
where
(9.22)
$$\mathbb{P}_{H,\tau}(t)f(t) = \mathbb{B}_{H,\tau}^* \mathbb{I}_{H,T}^* f(t).$$

Proof We only need to show (9.21) for the case when $f(t) = I_n^{H,T}(f_n(t))$, where $f_n \in \Theta_{H,T} \times \Theta_{H,T}^{\otimes n}$.
$$\mathbb{E}\left[\int_0^T I_n^{H,T}(f_n(s))dB_s^H\Big|\mathcal{F}_\tau\right] = \mathbb{E}\left[I_{n+1}^{H,T}(\tilde{f}_{n+1})\Big|\mathcal{F}_\tau\right],$$
where
$$\tilde{f}_{n+1}(t_1,\cdots,t_{n+1}) = \frac{1}{n+1}\Big[f_n(t_2,\cdots,t_{n+1};t_1)$$
$$+f_n(t_1,t_3,\cdots,t_{n+1};t_2)+\cdots+f_n(t_1,\cdots,t_n;t_{n+1})\Big].$$
From (9.12) it follows that
$$\mathbb{E}\left[\int_0^T I_n^{H,T}(f_n(t))dB_s^H|\mathcal{F}_\tau\right] = I_{n+1}^{H,\tau}(\mathbb{P}_{H,\tau}^{\otimes(n+1)}\tilde{f}_{n+1})$$
$$= \int_0^\tau \mathbb{P}_{H,\tau}(t)I_n^{H,\tau}(\mathbb{P}_{H,\tau}^{\otimes n}\tilde{f}_{n+1}(t))dB_t^H$$
$$= \int_0^\tau \mathbb{P}_{H,\tau}(t)I_n^{H,\tau}(\mathbb{P}_{H,\tau}^{\otimes n}f_{n+1}(t))dB_t^H$$
$$= \int_0^\tau \mathbb{P}_{H,\tau}(t)\mathbb{E}\left[f(t)|\mathcal{F}_\tau\right]dB_t^H.$$

This shows the theorem. $\qquad\square$

CHAPTER 10

Integration By Parts

First let us prove the Fubini type theorem.

THEOREM 10.1. *Let $(f(s,t)\,;\ s,t \in [0,T])$ be a stochastic process of the form*

$$f(s,t) = \sum_{i,j=1}^{n} p_{ij}(\tilde{f}_1, \cdots, \tilde{f}_n) f_i(s) f_j(t), \quad 0 \le s,t \le T,$$

where $p_{ij}, i,j = 1, \cdots, n$, are polynomials of n variables and $f_1, \cdots, f_n \in \mathbf{S}([0,T])$ and $\tilde{f}_k = \int_0^T f_k(t) dB_t^H$. Then
(a)

$$(10.1) \qquad \int_0^T \left[\int_0^T f(s,t) dB_s^H \right] dB_t^H = \int_0^T \left[\int_0^T f(s,t) dB_t^H \right] dB_s^H \,.$$

(b)

$$(10.2) \qquad \int_0^T \left[\int_0^T f(s,t) ds \right] dB_t^H = \int_0^T \left[\int_0^T f(s,t) dB_t^H \right] ds \,.$$

It is obvious that this theorem can be extended to more general stochastic processes.

THEOREM 10.2. *Let $F \in \mathbb{D}_{1,2}^H$ and let $(f(t), t \in [0,T])$ be (not necessarily adapted) stochastic process such that $f \in \mathbb{L}^{1,2}(\Omega, \mathcal{F}, P^H)$. Then*

$$(10.3) \qquad F \int_0^T f(t) dB_t^H = \int_0^T F f(t) dB_t^H + \int_0^T \left(\mathbb{D}_t^H F \right) f(t) dt \,.$$

Proof Let $G \in \mathbb{P}_H$. Then by Theorem 6.23 we have

$$\begin{aligned}
\mathbb{E} \left(\int_0^T F f(t) dB_t^H G \right) &= \int_0^T \mathbb{E} \left(F f(t) \mathbb{D}_t^H G \right) dt \\
&= \int_0^T \mathbb{E} \left\{ f(t) \left[\mathbb{D}_t^H (FG) - G \mathbb{D}_t^H F \right] \right\} dt \\
&= \int_0^T \mathbb{E} \left\{ f(t) \mathbb{D}_t^H (FG) - G f(t) \mathbb{D}_t^H F \right\} dt \\
&= \int_0^T \mathbb{E} f(t) \mathbb{D}_t^H (FG) dt - \int_0^T \mathbb{E} G f(t) \mathbb{D}_t^H F dt \\
&= \mathbb{E} \left\{ \left[F \int_0^T f(t) dB_t^H - \int_0^T f(t) \mathbb{D}_t^H F dt \right] G \right\} .
\end{aligned}$$

Since $G \in \mathbb{P}_H$ is arbitrary we obtain

$$\int_0^T Ff(t)dB_t^H = F\int_0^T f(t)dB_t^H - \int_0^T f(t)\mathbb{D}_t^H Fdt\,.$$

This completes the proof of the theorem. $\square$

The following theorem is an integration by parts formula.

THEOREM 10.3. *Let $T \in (0,\infty)$ and let $f_2(s)$ and $g_2(s)$ are in $\mathbb{D}_{1,2}^H$ and for $i = 1, 2$,*

(10.4) $$\mathbb{E}\left[\int_0^T |f_i(s)|^2 ds + \mathbb{E}\int_0^T |g_i(s)|^2 ds\right] < \infty\,.$$

Assume also that $\mathbb{D}_t^H f_2(s)$ and $\mathbb{D}_t^H g_2(s)$ are continuously differentiable with respect to $(s,t) \in [0,T]^2$ for P^H-almost all $\omega \in \Omega$. Suppose that $\mathbb{E}\int_0^T \int_0^T |\mathbb{D}_t^H f_2(s)|^2 < \infty$ and $\mathbb{E}\int_0^T \int_0^T |\mathbb{D}_t^H g_2(s)|^2 < \infty$. Denote

(10.5) $$F(t) = \int_0^t f_1(s)ds + \int_0^t f_2(s)dB_s^H\,, \quad t \in [0,T]$$

and

(10.6) $$G(t) = \int_0^t g_1(s)ds + \int_0^t g_2(s)dB_s^H\,, \quad t \in [0,T]\,.$$

Then

$$F(t)G(t) = \int_0^t F(s)g_1(s)ds + \int_0^t F(s)g_2(s)dB_s^H$$

$$+ \int_0^t G(s)f_1(s)ds + \int_0^t G(s)f_2(s)dB_s^H$$

(10.7) $$+ \int_0^t \mathbb{D}_s^H F(s)g_2(s)ds + \int_0^t \mathbb{D}_s^H G(s)f_2(s)ds\,.$$

Denote

$$dG(t) = g_1(t)dt + g_2(t)dB_t^H\,,$$

which means that

$$\int_0^t f(s)dG(s) = \int_0^t f(s)g_1(s)ds + \int_0^t f(s)g_2(s)dB_s^H\,.$$

The same notation will be applied to $dF(t)$.

REMARK 10.4. With these notations, Eq. (10.7) may be written formally as

(10.8) $d(F(t)G(t)) = F(t)dG(t) + G(t)dF(t) + \left[\mathbb{D}_t^H F(t)g_2(t) + \mathbb{D}_t^H G(t)f_2(t)\right]dt\,.$

Proof It suffices to show (10.6) for $t = T$. The general case is treated in a similar way except notational complexity. From (10.3) it follows that

$$
\begin{aligned}
F(T)G(T) &= G(T)\int_0^T f_1(t)dt + G(T)\int_0^T f_2(t)dB_t^H \\
&= \int_0^T f_1(t)G(T)dt + \int_0^T G(T)f_2(t)dB_t^H + \int_0^T \left[\mathbb{D}_t^H G(T)\right] f_2(t)dt \\
&= I_1 + I_2 + \int_0^T f_1(t)G(t)dt \\
&\quad + \int_0^T G(t)f_2(t)dB_t^H + \int_0^T \mathbb{D}_t^H G(T)f_2(t)dt \\
&= I_1 + I_2 + \int_0^T G(t)dF(t) + \int_0^T \mathbb{D}_t^H G(T)f_2(t)dt\,,
\end{aligned}
$$
(10.9)

where

$$
I_1 = \int_0^T f_1(t)\left[G(T) - G(t)\right]dt
$$
(10.10)

and

$$
I_2 = \int_0^T f_2(t)\left[G(T) - G(t)\right]dB_t^H\,.
$$
(10.11)

I_1 is calculated as follows.

$$
\begin{aligned}
I_1 &= \int_0^T f_1(t)\left[\int_t^T g_1(s)ds + \int_t^T g_2(s)dB_s^H\right]dt \\
&= \int_0^T \left[f_1(t)\int_t^T g_1(s)ds\right]dt + \int_0^T \left[f_1(t)\int_t^T g_2(s)dB_s^H\right]dt \\
&= \int_0^T \left[\int_t^T f_1(t)g_1(s)ds\right]dt + \int_0^T \left[\int_t^T f_1(t)g_2(s)dB_s^H\right]dt \\
&\quad + \int_0^T \int_t^T \mathbb{D}_s^H f_1(t)g_2(s)dsdt \\
&= \int_0^T \left[\int_0^s f_1(t)dt\right]g_1(s)ds + \int_0^T \left[\int_0^s f_1(t)dt\right]g_2(s)dB_s^H \\
&\quad + \int_0^T \int_t^T \mathbb{D}_s^H f_1(t)g_2(s)dsdt \\
&= \int_0^T \left[\int_0^s f_1(t)dt\right]dG(s) + \int_0^T \int_t^T \mathbb{D}_s^H f_1(t)g_2(s)dsdt\,.
\end{aligned}
$$
(10.12)

Now we compute I_2 as follows.

$$
\begin{aligned}
I_2 &= \int_0^T f_2(t) \left[\int_t^T g_1(s)ds + \int_t^T g_2(s)dB_s^H \right] dB_t^H \\
&= \int_0^T \left[\int_t^T f_2(t)g_1(s)ds \right] dB_t^H + \int_0^T \left[\int_t^T f_2(t)g_2(s)dB_s^H \right] dB_t^H \\
&\quad + \int_0^T \int_t^T \mathbb{D}_s^H f_2(t)g_2(s)ds dB_t^H \\
&= \int_0^T \int_0^s f_2(t)g_1(s)dB_t^H ds + \int_0^T \int_0^s f_2(t)g_2(s)dB_t^H dB_s^H \\
&\quad + \int_0^T \int_t^T \mathbb{D}_s^H f_2(t)g_2(s)ds dB_t^H \\
&= \int_0^T \left[\int_0^s f_2(t)dB_t^H \right] g_1(s)ds - \int_0^T \int_0^s \mathbb{D}_t^H g_1(s)f_2(t)dt ds \\
&\quad + \int_0^T \left[\int_0^s f_2(t)dB_t^H \right] g_2(s)dB_s^H - \int_0^T \int_0^s \mathbb{D}_t^H g_2(s)f_2(t)dt dB_s^H \\
&\quad + \int_0^T \int_t^T \mathbb{D}_s^H f_2(t)g_2(s)ds dB_t^H \\
&= \int_0^T \left[\int_0^s f_2(t)dB_t^H \right] dG(s) - \int_0^T \int_0^s \mathbb{D}_t^H g_1(s)f_2(t)dt ds \\
&\quad - \int_0^T \int_0^s \mathbb{D}_t^H g_2(s)f_2(t)dt dB_s^H + \int_0^T \int_t^T \mathbb{D}_s^H f_2(t)g_2(s)ds dB_t^H
\end{aligned}
$$

Therefore,

$$
\begin{aligned}
I_1 + I_2 &= \int_0^T F(s)dG(s) + \int_0^T \int_t^T \mathbb{D}_s^H f_1(t)g_2(s)ds dt + \int_0^T \int_t^T \mathbb{D}_s^H f_2(t)g_2(s)ds dB_t^H \\
&\quad - \int_0^T \int_0^s \mathbb{D}_t^H g_1(s)f_2(t)dt ds - \int_0^T \int_0^s \mathbb{D}_t^H g_2(s)f_2(t)dt dB_s^H \\
&= \int_0^T F(s)dG(s) + \int_0^T \int_0^s \mathbb{D}_s^H f_1(t)g_2(s)dt ds + \int_0^T \int_0^s \mathbb{D}_s^H f_2(t)g_2(s)dB_t^H ds \\
&\quad - \int_0^T \int_t^T \mathbb{D}_t^H g_1(s)f_2(t)ds dt - \int_0^T \int_t^T \mathbb{D}_t^H g_2(s)f_2(t)dB_s^H dt \\
&= \int_0^T F(s)dG(s) + \int_0^T \int_0^s \mathbb{D}_s^H f_1(t)g_2(s)dt ds \\
&\quad + \int_0^T g_2(s) \int_0^s \mathbb{D}_s^H f_2(t)dB_t^H ds - \int_0^T \int_0^s \mathbb{D}_s^H f_2(t)\mathbb{D}_t^H g_2(s)dt ds \\
&\quad - \int_0^T \int_t^T \mathbb{D}_t^H g_1(s)f_2(t)ds dt - \int_0^T \int_t^T \mathbb{D}_t^H g_2(s)f_2(t)dB_s^H dt
\end{aligned}
$$

Hence

$$
\begin{aligned}
I_1 + I_2 &= \int_0^T F(s)dG(s) - \int_0^T \int_0^s \mathbb{D}_s^H f_2(t)\mathbb{D}_t^H g_2(s)dtds \\
&\quad + \int_0^T \left[\int_0^s \mathbb{D}_s^H f_1(t)dt + \int_0^s \mathbb{D}_s^H f_2(t)dB_t^H \right] g_2(s)ds \\
&\quad - \int_0^T \left[\int_t^T \mathbb{D}_t^H g_1(s)ds + \int_t^T \mathbb{D}_t^H g_2(s)dB_s^H \right] f_2(t)dt \\
&\quad + \int_0^T \int_t^T \mathbb{D}_s^H f_2(t)\mathbb{D}_t^H g_2(s)dsdt \\
&= \int_0^T F(s)dG(s) + \int_0^T \left[\int_0^s \mathbb{D}_s^H f_1(t)dt + \int_0^s \mathbb{D}_s^H f_2(t)dB_t^H \right] g_2(s)ds \\
&\quad - \int_0^T \left[\int_t^T \mathbb{D}_t^H f_1(t)dt + \int_t^T \mathbb{D}_t^H g_2(s)dB_s^H \right] f_2(t)dt
\end{aligned}
$$

Notice that for any $h \in \mathbf{S}$ and $-\infty < a < b < \infty$,

$$
\mathbb{D}_t^H \int_0^T h(s)dB_s^H = \int_0^T \mathbb{D}_t^H h(s)dB_s^H + \mathbf{B}_{H,T}h(t)\,,
$$

where

$$
\mathbf{B}_{H,T} = \mathbb{I}_{H,T}\mathbb{I}_{H,T}^*\,.
$$

Therefore,

$$
\begin{aligned}
I_1 + I_2 &= \int_0^T F(s)dG(s) + \int_0^T \left\{ \mathbb{D}_s^H F(s) - \mathbf{B}_{H,T}(u)\left(\chi_{[0,s]}(u)f_2(u) \right)\Big|_{u=s} \right\} g_2(s)ds \\
&\quad - \int_0^T \left\{ \mathbb{D}_s^H \left(G(T) - G(s) \right) - \mathbf{B}_{H,T}\left(\chi_{[s,T]}(u)g_2(u) \right)\Big|_{u=s} \right\} f_2(s)ds \\
&= \int_0^T F(s)dG(s) + \int_0^T \mathbb{D}_s^H F(s)g_2(s)ds \\
&\quad - \int_0^T \mathbb{D}_s^H \left[G(T) - G(s) \right] f_2(s)ds + I_3\,,
\end{aligned}
$$

where

$$
\begin{aligned}
I_3 &= \int_0^T \left(\mathbf{B}_{H,T}\chi_{[s,T]}(u)g_2(u) \right)\Big|_{u=s} f_2(s)ds \\
&\quad - \int_0^T \left(\mathbf{B}_{H,T}\chi_{[0,s]}(u)f_2(u) \right)\Big|_{u=s} g_2(s)ds\,.
\end{aligned}
\tag{10.13}
$$

Now we compute I_3. Write $I_3 = J_1 - J_2$, where

$$
J_1 := \int_0^T \mathbb{I}_{H,T}(t)\mathbb{I}_{H,T}^*(t)\left(\chi_{[s,T]}(t)g_2(t) \right)\Big|_{t=s} f_2(s)ds
$$

and

$$
J_2 := \int_0^T \mathbb{I}_{H,T}(t)\mathbb{I}_{H,T}^*(t)\left(\chi_{[0,s]}(t)f_2(t) \right)\Big|_{t=s} g_2(s)ds\,.
$$

If $0 < H < 1/2$ then it follows from (5.30) that

$$
\begin{aligned}
J_1 &= H \int_0^T (T-s)^{2H-1} f_2(s) ds g_2(T) \\
&\quad + H \int_0^T \int_s^T |s-u|^{2H-1} \mathrm{sign}\,(s-u) g_2'(u) du f_2(s) ds \\
&= H \int_0^T (t-s)^{2H-1} f_2(s) ds g_2(T) \\
&\quad + H \int_0^T \int_0^u |s-u|^{2H-1} \mathrm{sign}\,(s-u) f_2(s) ds g_2'(u) du \\
&= H \int_0^T (t-s)^{2H-1} f_2(s) ds g_2(T) + \frac{1}{2} \int_0^T \left\{ |s-u|^{2H} f_2(s) \big|_0^u \right\} g_2'(u) du \\
&\quad - \frac{1}{2} \int_0^T \int_0^u |s-u|^{2H} g_2'(u) f_2'(s) ds du \\
&= H \int_0^T (t-s)^{2H-1} f_2(s) ds g_2(T) - \frac{1}{2} \int_0^T u^{2H} f_2(0) g_2'(u) du \\
&\quad - \frac{1}{2} \int_0^T \int_0^u |s-u|^{2H} g_2'(u) f_2'(s) ds du \\
&= H \int_0^T (T-s)^{2H-1} f_2(s) ds g_2(T) - \frac{1}{2} f_2(0) T^{2H} g_2(T)
\end{aligned}
$$

$$
\tag{10.14}
\quad + H \int_0^T u^{2H-1} g_2(u) du f_2(0) - \frac{1}{2} \int_0^T \int_0^u |s-u|^{2H} g_2'(u) f_2'(s) ds du \,.
$$

Similarly,

$$
\begin{aligned}
J_2 &= H \int_0^T s^{2H-1} g_2(s) ds f_2(0) \\
&\quad + H \int_0^T \int_u^T |s-u|^{2H-1} \mathrm{sign}\,(s-u) g_2(s) ds f_2'(u) du \\
&= H \int_0^T s^{2H-1} g_2(s) ds f_2(0) + \frac{1}{2} \int_0^T |T-u|^{2H} f_2'(u) du g_2(T) \\
&\quad - \frac{1}{2} \int_0^T \int_0^s |s-u|^{2H} g_2'(s) ds f_2'(u) du \\
&= H \int_0^T s^{2H-1} g_2(s) ds f_2(0) - \frac{1}{2} g_2(T) T^{2H} f_2(0) \\
&\quad + H \int_0^T (T-u)^{2H-1} f_2(u) du g_2(T)
\end{aligned}
$$

$$
\tag{10.15}
\quad - \frac{1}{2} \int_0^T \int_0^u |s-u|^{2H} g_2'(u) f_2'(s) ds du \,.
$$

Combining Eqs. (10.13)-(10.15), we obtain that

$$
\tag{10.16}
I_3 = J_1 - J_2 = 0 \quad \text{if} \quad 0 < H < 1/2 \,.
$$

If $1/2 < H < 1$, then from (5.29) it follows that

$$
\begin{aligned}
J_1 &= H(2H-1) \int_0^T \int_s^T |s-u|^{2H-2} g_2(u) du \, f_2(s) ds \\
&= H(2H-1) \int_0^T \int_0^u |s-u|^{2H-2} f_2(s) ds \, g_2(u) du
\end{aligned}
$$

and

$$
J_2 = H(2H-1) \int_0^T \int_0^s |s-u|^{2H-2} f_2(u) du \, g_2(s) ds \,.
$$

Hence

(10.17) $$ I_3 = J_1 - J_2 = 0 \quad \text{if} \quad 1/2 < H < 1 \,. $$

Thus for all $0 < H < 1$ we have

$$
I_1 + I_2 = \int_0^T F(s) dG(s) + \int_0^T \mathbb{D}_s^H F(s) g_2(s) ds - \int_0^T \mathbb{D}_s^H \left[G(T) - G(s) \right] f_2(s) ds \,.
$$

Substituting $I_1 + I_2$ into Eq. (10.8) by the above identity, we conclude that

$$
F(T)G(T) = \int_0^T F(t) dG(t) + \int_0^T G(t) dF(t) + \int_0^T \left[\mathbb{D}_t^H F(t) g_2(t) + \mathbb{D}_t^H G(t) f_2(t) \right] dt \,.
$$

This completes the proof of Eq. (10.6). $\qquad\qquad\qquad\qquad\qquad\qquad\square$

CHAPTER 11

Composition (Itô Formula)

First we shall obtain an Itô type formula for $F(\int_0^T f_s dB_s^H)$ for general deterministic f and general $H \in (0,1)$.

THEOREM 11.1. *Let $0 < H < 1$ and let $f \in \Theta_{H,T} \cap L^2([0,T])$ be an deterministic function. Denote $f_t(s) = f(s)\chi_{[0,t]}(s)$, $0 \le s \le t \le T$. Suppose that $f_t \in \Theta_{H,t}$ and $\|f_t\|_{\Theta_{H,t}}$ is continuously differentiable as a function of $t \in [0,T]$. Denote*

$$(11.1) \qquad X_t = X_0 + \int_0^t g_s ds + \int_0^t f_s dB_s^H , \quad 0 \le t \le T ,$$

where X_0 is a constant, g is deterministic with $\int_0^T |g_s| ds < \infty$. Let F be an entire function of order less than 2. Namely,

$$M_f(r) := \sup_{|z|=r} |f(z)| < Ce^{Ar^K} \quad for\ all \quad r ,$$

where K is a positive number less than 2 and C is a constant. Then

$$
\begin{aligned}
F(t, X_t) \;&=\; F(0, X_0) + \int_0^t \frac{\partial F}{\partial s}(s, X_s) ds + \int_0^t \frac{\partial F}{\partial x}(s, X_s) dX_s \\
(11.2) \qquad &+ \frac{1}{2} \int_0^t \frac{\partial^2 F}{\partial x^2}(s, X_s) \left[\frac{d}{ds} \|f_s\|_{\Theta_{H,s}}^2 \right] ds , \quad 0 \le t \le T .
\end{aligned}
$$

REMARK 11.2. The stochastic integral in (11.2) is in the algebraically integrable sense.

Proof We shall prove the above theorem for $t = T$ and $X_0 = g = 0$. We will write f_t as $f I_{[0,t]}$ if there is no ambiguity. Consider

$$
\begin{aligned}
&\exp\left(\xi \int_0^t f(s) dB_s^H - \frac{\xi^2}{2} \|f\|_{\Theta_{H,t}}^2 \right) \\
=\; &\exp\left(\xi \int_0^T f_t(s) dB_s^H - \frac{\xi^2}{2} \|f_t\|_{\Theta_{H,T}}^2 \right) \\
=\; &\sum_{n=0}^{\infty} \frac{\xi^n}{n!} I_n^{H,T} \left(f_t^{\otimes n} \right) .
\end{aligned}
$$

Note that the symmetrization of $f_t^{\otimes n} f(t)$ is $\dfrac{1}{n+1} f^{\otimes(n+1)}$. Thus we obtain

$$\int_0^T f(t) I_n^{H,T}(f_t^{\otimes n}) dB_t^H = \frac{1}{n+1} I_{n+1}^{H,T}(f^{\otimes(n+1)}) .$$

From this equality it is easy to see that $\exp\left(\xi \int_0^t f(s)dB_s^H - \frac{\xi^2}{2}\|f\|_{\Theta_{H,t}}^2\right) f(t)$ is algebraically integrable and

$$\int_0^T f(t)\exp\left(\xi \int_0^t f(s)dB_s^H - \frac{\xi^2}{2}\|f\|_{\Theta_{H,t}}^2\right)dB_t^H$$

$$= \sum_{n=0}^{\infty} \frac{\xi^n}{n!}\int_0^T I_n^{H,T}\left(f_t^{\otimes n}\right)f(t)dB_t^H$$

$$= \sum_{n=0}^{\infty} \frac{\xi^n}{(n+1)!}I_{n+1}^{H,T}\left(f^{\otimes(n+1)}\right)$$

$$= \frac{1}{\xi}\left[\exp\left(\xi \int_0^T f(s)dB_s^H - \frac{\xi^2}{2}\|f\|_{\Theta_{H,T}}^2\right) - 1\right].$$

Thus

(11.3)
$$\int_0^T f(t)\exp\left(\xi \int_0^t f(s)dB_s^H - \frac{\xi^2}{2}\|f\|_{\Theta_{H,t}}^2\right)dB_t^H$$

$$= \frac{1}{\xi}\left[\exp\left(\xi \int_0^T f(s)dB_s^H - \frac{\xi^2}{2}\|f\|_{\Theta_{H,T}}^2\right) - 1\right].$$

But

(11.4) $\exp\left(\frac{\xi^2}{2}\|f\|_{\Theta_{H,t}}^2\right) = \exp\left(\frac{\xi^2}{2}\|f\|_{\Theta_{H,T}}^2\right) - \int_t^T \exp\left(\frac{\xi^2}{2}\|f\|_{\Theta_{H,u}}^2\right)\xi^2 H_u du,$

where $H_u = \frac{1}{2}\frac{d}{du}\|f\|_{\Theta_{H,u}}^2$. Denote

$$\mathbf{E}(X) := \exp(X - \frac{1}{2}\mathbf{E}(X^2)) \quad \text{when } X \text{ is centered Gaussian random variable}.$$

Thus from (11.3)-(11.4) it follows that

$$\int_0^T f(t)\exp\left(\xi \int_0^t f(s)dB_s^H\right)dB_t^H$$

$$= \int_0^T \mathbf{E}\left(\xi \int_0^t f(s)dB_s^H\right)\exp\left(\frac{\xi^2}{2}\|f\|_{\Theta_{H,t}}^2\right)f(t)dB_t^H$$

$$= \int_0^T f(t)\mathbf{E}\left(\xi \int_0^t f(s)dB_s^H\right)dB_t^H \exp\left(\frac{\xi^2}{2}\|f\|_{\Theta_{H,T}}^2\right)$$

$$\quad -\xi^2\int_0^T\left[\int_t^T \exp\left(\frac{\xi^2}{2}\|f\|_{\Theta_{H,u}}^2\right)H_u du\right]\mathbf{E}\left(\xi \int_0^t f(s)dB_s^H\right)f(t)dB_t^H$$

$$= \frac{1}{\xi}\left[\mathbf{E}\left(\xi \int_0^T f(t)dB_t^H\right) - 1\right]\exp\left(\frac{\xi^2}{2}\|f\|_{\Theta_{H,T}}^2\right)$$

$$\quad -\xi^2\int_0^T\left[\int_0^u \mathbf{E}\left(\xi \int_0^t f(s)dB_s^H\right)f(t)dB_t^H\right]\exp\left(\frac{\xi^2}{2}\|f\|_{\Theta_{H,u}}^2\right)H_u du$$

$$= \frac{1}{\xi}\left[\exp\left(\xi \int_0^T f(t)dB_t^H\right) - \exp\left(\frac{\xi^2}{2}\|f\|_{\Theta_{H,T}}^2\right)\right]$$

$$\quad -\xi\int_0^T\left[\exp\left(\xi \int_0^u f(s)dB_s^H\right) - \exp\left(\frac{\xi^2}{2}\|f\|_{\Theta_{H,u}}^2\right)\right]H_u du.$$

Therefore

$$(11.5) \quad \exp\left(\xi \int_0^T f(t)dB_t^H\right)$$

$$= \exp\left(\frac{\xi^2}{2}\|f\|_{\Theta_{H,T}}^2\right) + \xi \int_0^T \exp\left(\xi \int_0^t f(s)dB_s^H\right) f(t)dB_t^H$$

$$+\xi^2 \int_0^T \exp\left(\xi \int_0^u f(s)dB_s^H\right) H_u du - \xi^2 \int_0^T \exp\left(\frac{\xi^2}{2}\|f\|_{\Theta_{H,u}}^2\right) H_u du$$

$$= 1 + \xi \int_0^T \exp\left(\xi \int_0^t f(s)dB_s^H\right) f(t)dB_t^H$$

$$+\xi^2 \int_0^T \exp\left(\xi \int_0^u f(s)dB_s^H\right) H_u du.$$

If we denote $F(x) = \exp(\xi x)$, where ξ is a complex number, then the above identity can be written as

$$F\left(\int_0^T f(t)dB_t^H\right) = F(0) + \int_0^T F'\left(\int_0^t f(s)dB_s^H\right) f(t)dB_t^H$$

$$(11.6) \qquad\qquad +\frac{1}{2}\int_0^T F''\left(\int_0^t f(s)dB_s^H\right) \left[\frac{d}{dt}\|f\|_{\Theta_{H,t}}^2\right] dt.$$

Since (11.5) is true for all complex number ξ, we see that (11.6) holds for polynomial $F(x) = x^n$. Namely,

$$\left(\int_0^T f(t)dB_t^H\right)^n = n\int_0^T \left(\int_0^t f(s)dB_s^H\right)^{n-1} f(t)dB_t^H$$

$$(11.7) \qquad\qquad +\frac{n(n-1)}{2}\int_0^T \left(\int_0^t f(s)dB_s^H\right)^{n-2} \left[\frac{d}{dt}\|f\|_{\Theta_{H,t}}^2\right] dt.$$

Let $F(x) = \sum_{n=0}^\infty a_n x^n$, $x \in \mathbb{R}$, be an entire function of finite order $K < 2$. Denote $F_k(x) = \sum_{n=0}^k a_n x^n$, $x \in \mathbb{R}$. It is easy to see that (11.5) holds for F_k, $k = 1, 2, \cdots$. Then it is well-known (see for example [**73**]) that

$$|a_n| \leq C^n n^{-n/K}.$$

We shall use the following fact: If X is Gaussian random variable with mean 0 and variance $\mathbb{E}(X^2)$. Then

$$\mathbb{E}(X^n) = \begin{cases} 0 & \text{if n is an odd number} \\ \frac{n!}{2^{n/2}(n/2)!} \left[\mathbb{E}(X^2)\right]^{n/2} & \text{if n is an even number} \end{cases}$$

From this we see that if n is even number, then

$$\mathbb{E}\left|\int_0^t f(s)dB_s^H\right|^n = \frac{n!}{2^{n/2}(n/2)!}\|f\|_{\Theta_{H,t}}^n.$$

If n is odd, then

$$
\mathbb{E}\left|\int_0^t f(s)dB_s^H\right|^n \leq \left(\mathbb{E}\left(\int_0^t f(s)dB_s^H\right)^{n+1}\right)^{1/2}\left(\mathbb{E}\left(\int_0^t f(s)dB_s^H\right)^{n-1}\right)^{1/2}
$$

$$
= \left(\frac{(n+1)!}{2^{(n+1)/2}((n+1)/2)!} \cdot \frac{(n-1)!}{2^{(n-1)/2}((n-1)/2)!}\right)^{/2}\|f\|_{\Theta_{H,t}}^n.
$$

Therefore we can conclude that there is a constant C such that

$$
\mathbb{E}\left|\int_0^t f(s)dB_s^H\right|^n = C^n([n/2]!)\|f\|_{\Theta_{H,t}}^n,
$$

where $[x]$ denotes the greatest integer less or equal to x. From (11.7) we obtain that

$$
\left|\int_0^T\left(\int_0^t f(s)dB_s^H\right)^{n-1}f(t)dB_t^H\right| \leq C^n([n/2]!).
$$

Thus we obtain

$$
\mathbb{E}\left|\int_0^T F'\left(\int_0^t f(s)dB_s^H\right)f(t)dB_t^H - \int_0^T F_k'\left(\int_0^t f(s)dB_s^H\right)f(t)dB_t^H\right|
$$

$$
\leq \sum_{n=k+1}^\infty |a_n|\mathbb{E}\left|\int_0^T\left(\int_0^t f(s)dB_s^H\right)^{n-1}f(t)dB_t^H\right|
$$

$$
\leq \sum_{n=k+1}^\infty |a_n|C^n([n/2]!) \leq \sum_{n=k+1}^\infty C^n n^{-n/K}([n/2]!) \to 0 \quad \text{when} \quad k \to \infty.
$$

Hence there is a subsequence such that

$$
\int_0^T F_k'\left(\int_0^t f(s)dB_s^H\right)f(t)dB_t^H \quad \text{converegs to} \quad \int_0^T F'\left(\int_0^t f(s)dB_s^H\right)f(t)dB_t^H
$$

almost surely. The other term can be treated in similar way. This proves the theorem.   □

REMARK 11.3. Let $F(x) = \int_{\mathbb{R}} e^{ix\xi}\mu(d\xi)$, where μ is a measure on $\mathbb{R}$ such that $\int_{\mathbb{R}}(1+|x|^2)\mu(dx) < \infty$. Then it is easy to check (11.2) holds for F.

The following corollary is interesting and is independently obtained by [4], [14] by using the white noise calculus started in [35] and [58].

COROLLARY 11.4. Let $0 < H < 1$ and let F satisfy the conditions of Theorem 11.1. Then

$$
F(t, B_t^H) = F(0,0) + \int_0^t \frac{\partial F}{\partial s}(s, B_s^H)ds + \int_0^t \frac{\partial F}{\partial x}(s, B_s^H)dB_s^H
$$

(11.8)
$$
+ H\int_0^t \frac{\partial^2 F}{\partial x^2}(s, B_s^H)s^{2H-1}ds, \quad 0 \leq t \leq T.
$$

Proof Let $f = 1$. Then $\|f\|_{\Theta_{H,t}} = t^H$. It is easy to see that it is differentiable and $\frac{d}{dt}\|f\|_{\Theta_{H,t}} = Ht^{H-1}$. Thus the corollary is a direct consequence of Theorem 11.1.
□

When the Hurst parameter H is greater than $1/2$ an Itô formula for more general integral process is proved in [**35**]. In that paper, a Malliavin derivative D^ϕ is introduced as follows

$$D_s^\phi F \;=\; H(2H-1)\int_0^T |s-r|^{2H-2} D_r^H F dr \,.$$

Note that D^ϕ is the same as $\mathbb{D}^H$ in our notation here for $H > 1/2$. To be consistent with our notation we shall use $\mathbb{D}^H$ to denote D^ϕ. The following theorem is the Theorem 4.3 of [**35**].

THEOREM 11.5. *Let $\eta_t = \int_0^t F_u dB_u^H$, where $(F_u, 0 \le u \le T)$ is a stochastic process in $L_H^2([0,T])$. Assume that there is an $\alpha > 1 - H$ such that*

$$\mathbb{E}\,|F_u - F_v|^2 \le C|u-v|^{2\alpha} \,.$$

where $|u-v| \le \delta$ for some $\delta > 0$ and

$$\lim_{0 \le u,v \le t, |u-v| \to 0} \mathbb{E}\,|\mathbb{D}_u^H (F_u - F_v)|^2 = 0 \,.$$

Let $f : \mathbb{R}_+ \times \mathbb{R} \to \mathbb{R}$ be a function having the first continuous derivative in its first variable and the second continuous derivative in its second variable. Assume that these derivatives are bounded. Moreover, it is assumed that $\mathbb{E}\int_0^T |F_s \mathbb{D}_s^H \eta_s| ds < \infty$ and $(f'(s,\eta_s) F_s, s \in [0,T])$ is in $L_H^2([0,T])$. Then for $0 \le t \le T$,

$$
\begin{aligned}
f(t,\eta_t) \;=\;& f(0,0) + \int_0^t \frac{\partial f}{\partial s}(s,\eta_s)ds + \int_0^t \frac{\partial f}{\partial x}(s,\eta_s)F_s dB_s^H \\
& + \int_0^t \frac{\partial^2 f}{\partial x^2}(s,\eta_s)F_s \mathbb{D}_s^H \eta_s ds \quad a.s.
\end{aligned}
$$

(11.9)

By the property of $\mathbb{D}^H$ we have

$$D_r^H \eta_s = \int_0^s D_r^H F_\theta dB_\theta^H + F_r \,.$$

If $H > 1/2$, then

$$
\begin{aligned}
\mathbb{D}_s^H \eta_s \;=\;& H(2H-1)\int_0^s \int_0^s |s-r|^{2H-2} D_r^H F_\theta dr dB_\theta^H \\
& + H(2H-1)\int_0^s |s-r|^{2H-2} F_r dr \,.
\end{aligned}
$$

(11.10)

Therefore the Itô formula can also take the following form

THEOREM 11.6. *Let η and f be defined as in Theorem 11.3 and let the conditions of the theorem be satisfied. Then*

$$
\begin{aligned}
f(t,\eta_t) =\;& f(0,0) + \int_0^t \frac{\partial f}{\partial s}(s,\eta_s)ds + \int_0^t \frac{\partial f}{\partial x}(s,\eta_s)F_s dB_s^H \\
& + H(2H-1)\int_0^t \frac{\partial^2 f}{\partial x^2}(s,\eta_s)F_s \left[\int_0^s \int_0^s |s-r|^{2H-2} D_r^H F_\theta dr dB_\theta^H\right] F_s ds \\
& + H(2H-1)\int_0^t \int_0^s \frac{\partial^2 f}{\partial x^2}(s,\eta_s)F_s |s-r|^{2H-2} F_s F_r dr ds \,. \quad a.s.
\end{aligned}
$$

(11.11)

In [**3**], Alòs, Mazet, and Nualart defines stochastic integral for a general Gaussian process

$$B_t^K = \int_0^t K(t,s)dB_s\,,$$

where $K(\cdot,s)$ has bounded variation on any interval (u,T), $u > s$ and

$$\int_0^T \left(\int_s^T \|B_t^K - B_s^K\|_{L^2(\Omega,\mathcal{F},P)} |K|(dt,s) \right)^2 ds < \infty\,.$$

REMARK 11.7. (a) As noticed in [**3**] this condition is satisfied only if $H > 1/4$.
(b) We use different notations to denote the Brownian motion and B^K than those in [**3**]. B and B^K in this paper are denoted by W and B respectively in [**3**].

Denote

$$(Kh)_t = (Kh)(t) = \int_0^t K(t,s)f(s)ds$$

and

$$(K^*h)(s) = h(s)K(T,s) + \int_s^T [\phi(t) - \phi(s)]\, K(dt,s)\,.$$

Then K^* is the adjoint operator of K. Namely,

$$\int_0^T (K^*g)(t)f(t)dt = \int_0^T (Kf)(t)g(t)dt\,.$$

The stochastic integral in [**3**] is defined as

$$\int_0^T u_s dB_s^K = \int_0^T (K^*u)_s dB_s\,,$$

if the above last integral exists in the sense of Skorohod type integral.

THEOREM 11.8. *Let F be a twice continuously differentiable function on $\mathbb{R}$ with bounded derivatives. Suppose that $(B_t^K, 0 \le t \le T)$ is a zero mean continuous Gaussian process whose covariance function $R(t,s) = \mathbb{E}\left(B_t^K B_s^K\right)$ is of the form*

$$R(t,s) = \int_0^{t \wedge s} K(t,r)K(s,r)dr\,,$$

where $K(t,s)$ satisfies the following assumptions.
(i) $K(t,s)$ is differentiable in t in $\{0 < s < t < T\}$, and both K and $\frac{\partial K}{\partial t}$ are continuous in $\{0 < s < t < T\}$.
(ii) There is $\alpha < 1/4$ and $0 < C < \infty$ such that

(11.12) $$|\frac{\partial K}{\partial t}(t,s)| \le C(t-s)^{-\alpha-1}\,, \qquad \text{and}$$

(iii)

(11.13) $$\int_0^t K(t,u)^2 du < C(t-s)^{1-2\alpha}\,.$$

Let $u = \{u_t, t \in [0,T]\}$ *be an adapted process in the space $\mathbb{D}^{2,2}(\Omega, \mathcal{F}, P)$ satisfying the following conditions*

(iv) The process u, $D_r u$ are λ-Hölder continuous in the norm of the space $\mathbb{D}^{1,4}(\Omega, \mathcal{F}, P)$ for some $\lambda > \alpha$, and the function

$$(11.14) \qquad \gamma_r = \sup_{0 \le s \le T} \|D_r u_s\|_{1,4} + \sup_{0 \le s \le t \le T} \frac{\|D_r u_t - D_r u_s\|_{1,4}}{t - s}$$

satisfies $\int_0^T \gamma_r^p dr < \infty$ for some $p > \frac{2}{1-4\alpha}$ and

$$(11.15) \qquad \sup_{\varepsilon > 0} \mathbb{E}^P \int_0^T \left| \frac{\partial}{\partial s} \int_0^s (K_s^{*,\varepsilon} u)_r \, dr \right|^2 ds < \infty.$$

Then $X_t = \int_0^t u_s dB_s^K$ exists and

$$
\begin{aligned}
F(X_t) &= F(0) + \int_0^t F'(X_s) u_s dB_s^K \\
&\quad + \int_0^t F''(X_s) u_s \left(\int_0^s \frac{\partial K}{\partial s}(s,r) \int_0^s D_r (K_s^* u)_\theta dB_\theta \right) ds \\
&\quad + \frac{1}{2} \int_0^t F''(X_s) \frac{\partial}{\partial s} \left(\int_0^s (K_s^* u)_r^2 dr \right) ds.
\end{aligned}
$$
(11.16)

It is easy to check that if $K(t,s) = Z_H(t,s)$, then (ii) implies that $H > 1/4$. The condition $H > 1/4$ can also be seen heuristically from the definition of stochastic integral in the sense of [**3**]:

$$\int_0^t F'(X_s) u_s dB_s^H = \int_0^t (K_s^* F'(X.) u.)_s dB_s.$$

Roughly speaking (formally), K_s^* is a "differential operator" (or "adjoint differential operator") of order $\frac{1}{2} - H$. This operator transforms a Hölder continuous function of order $\frac{1}{2} - H$ to continuous function. It is also clear that B_s^H is a Hölder continuous functions (as a function of s) of order H. If we require that the form $F'(X_s) u_s$ contains form of function of B_s, then we need $H > \frac{1}{2} - H$, which is $H > 1/4$.

It appears that the two formulas (11.16) and (11.9) are different even in the case $H > 1/2$. We shall explain that they are the same formula when $H > 1/2$. On the other hand, if we take $K(t,s) = Z_H(t,s)$, then $(K_s^* u)_r = \mathbb{I}_{H,s} u(r)$. Note that D_s^H defined in [**35**] or in this paper satisfies $D_s = (\mathbb{I}_{H,s}^* D^H)_s$. Therefore the third term of (11.6) is

$$(11.17) \quad \int_0^s \frac{\partial K}{\partial s}(s,r) \int_0^s D_r (K_s^* u)_\theta dB_\theta dr = \int_0^s \mathbb{I}_{H,s} (\mathbb{I}_{H,s}^* D_s^H) u_\theta dB_\theta^H.$$

It is clear that

$$(K_s K_s^* u)_r = H(2H-1) \int_0^s |r - \xi|^{2H-2} u(\xi) d\xi.$$

Hence

$$
\begin{aligned}
\frac{\partial}{\partial s} \left(\int_0^s (K_s^* u)_r^2 dr \right) ds &- \frac{\partial}{\partial s} \int_0^s u(r)(K_s K_s^*) u(r) dr \\
&= H(2H-1) \frac{\partial}{\partial s} \int_0^s u(r) \int_0^s |r - \xi|^{2H-2} u(\xi) d\xi dr \\
&= 2H(2H-1) u(s) \int_0^s |s - r|^{2H-2} u(r) dr \\
&= 2u(s) \mathbb{I}_{H,s} \mathbb{I}_{H,s}^* u(s).
\end{aligned}
$$
(11.18)

Summarizing above we obtain

THEOREM 11.9. *Let the condition of Theorem 11.5 hold for* $K = Z_H(t,s)$, *where* $H > 1/2$. *Then*

$$
\begin{aligned}
F(X_t) \;=\; & F(0) + \int_0^t F'(X_s)u_s dB_s^K \\
& + \int_0^t F''(X_s)u_s \left(\int_0^s \mathbb{I}_{H,s}(\mathbb{I}_{H,s}^* D_s^H)u_\theta dB_\theta^H \right) ds \\
& + \int_0^t F''(X_s)\left(u(s)\mathbb{I}_{H,s}\mathbb{I}_{H,s}^* u(s) \right) ds\,.
\end{aligned}
$$
(11.19)

Using the fact that $\mathbb{I}_{H,s}\mathbb{I}_{H,s}^* u = 2H(2H-1)\int_0^s |s-r|^{2H-2}u(r)dr$ we see that the Itô formula established in [**35**] (e.g. (11.9)) and that of [**3**] (e.g. (11.16)) are the same when $H > 1/2$.

We are going to discuss the condition on the differentiability of $\|f\|_{\Theta_{H,t}}^2$ in the Theorem 11.1.

Let $H > 1/2$. By the definition of the space $\Theta_{H,t}$ and (5.28) we have

$$
\begin{aligned}
\|f\|_{\Theta_{H,t}}^2 \;=\; & \int_0^t |\mathbb{I}_{H,t}f(v)|^2 dv \\
=\; & \int_0^t f(v)\mathbb{I}_{H,t}\mathbb{I}_{H,t}^* f(v)dv \\
=\; & H(2H-1)\int_0^t \int_0^t |v-u|^{2H-2}f(u)f(v)dudv\,.
\end{aligned}
$$

If f is continuous on $[0,\,T]$, then it is easy to see that $\|f\|_{\Theta_{H,t}}^2$ is differentiable and

$$
\frac{d}{dt}\|f\|_{\Theta_{H,t}}^2 = 2H(2H-1)f(t)\int_0^t |t-u|^{2H-2}f(u)du\,, 0 \le t \le T\,.
$$
(11.20)

Let $H < 1/2$ and let f be continuously differentiable. Then by the definition of the space $\Theta_{H,t}$ and (5.30) we have

$$
\begin{aligned}
\|f\|_{\Theta_{H,t}}^2 \;=\; & \int_0^t |\mathbb{I}_{H,t}f(v)|^2 dv \\
=\; & \int_0^t f(v)\mathbb{I}_{H,t}\mathbb{I}_{H,t}^* f(v)dv \\
=\; & \int_0^t f(v)\Big[Hv^{2H-1}f(0) + H|t-v|^{2H-1}f(t) \\
& \qquad\qquad +H\int_0^t |v-u|^{2H-1}\mathrm{sign}\,(v-u)f'(u)du \Big] dv \\
=\; & Hf(0)\int_0^t v^{2H-1}f(v)dv + Hf(t)\int_0^t |t-v|^{2H-1}f(v)dv \\
& +H\int_0^t \int_0^t |v-u|^{2H-1}\mathrm{sign}\,(v-u)f'(u)f(v)dudv\,.
\end{aligned}
$$

Making substitution $u = t\xi$ we have

$$\frac{d}{dt}\int_0^t (t-u)^{2H-1}f(u)du$$

$$= \frac{d}{dt}\left(t^{2H}\int_0^1 (1-\xi)^{2H-1}f(t\xi)d\xi\right)$$

$$= 2Ht^{2H-1}\int_0^1 (1-\xi)^{2H-1}f(t\xi)d\xi + t^{2H}\int_0^1 (1-\xi)^{2H-1}\xi f'(t\xi)d\xi\,.$$

Therefore

$$\frac{d}{dt}\|f\|_{\Theta_{H,t}}^2 = Ht^{2H-1}f(0)f(t) + Hf'(t)\int_0^t |t-v|^{2H-1}f(v)dv$$

$$+2H^2 t^{2H-1}f(t)\int_0^1 (1-\xi)^{2H-1}f(t\xi)d\xi$$

$$+Ht^{2H}f(t)\int_0^1 (1-\xi)^{2H-1}\xi f'(t\xi)d\xi$$

$$+Hf(t)\int_0^t (t-u)^{2H-1}f'(u)du - Hf'(t)\int_0^t (t-v)^{2H-1}f(v)dv$$

$$= Ht^{2H-1}f(0)f(t) + 2H^2 t^{2H-1}f(t)\int_0^1 (1-\xi)^{2H-1}f(t\xi)d\xi$$

$$+Ht^{2H}f(t)\int_0^1 (1-\xi)^{2H-1}\xi f'(t\xi)d\xi$$

(11.21) $$+Hf(t)\int_0^t (t-u)^{2H-1}f'(u)du\,.$$

Thus as a consequence of Theorem 1.1, we have

COROLLARY 11.10. Let $0 < H < 1$ and let $f \in \Theta_{H,T} \cap L^2([0,T])$ be an deterministic function. Denote $f_t(s) = f(s)\chi_{[0,t]}(s)$, $0 \le s \le t \le T$. Let f be continuous over $[0,T]$ if $H > 1/2$ and let f be continuously differentiable over $[0,T]$ if $H < 1/2$. Denote

(11.22) $$X_t = X_0 + \int_0^t g_s ds + \int_0^t f_s dB_s^H\,, \quad 0 \le t \le T,$$

where X_0 is a constant, g is deterministic with $\int_0^T |g_s|ds < \infty$. Let F be an entire function of order less than 2. Then

$$F(t,X_t) = F(0,X_0) + \int_0^t \frac{\partial F}{\partial s}(s,X_s)ds + \int_0^t \frac{\partial F}{\partial x}(s,X_s)dX_s$$

(11.23) $$+\frac{1}{2}\int_0^t \frac{\partial^2 F}{\partial x^2}(s,X_s)\left[\frac{d}{ds}\|f_s\|_{\Theta_{H,s}}^2\right]ds\,, \quad 0 \le t \le T,$$

where $\frac{d}{ds}\|f_s\|_{\Theta_{H,s}}$ is given by (11.20) or (11.21)

REMARK 11.11. Other types of stochastic integral and the corresponding Itô formulas have been discussed in the literature. See for example, [24], [37], [42], [43].

CHAPTER 12

Clark Type Representation

In the sBm case it is well-known that for any square integrable Wiener functional F, there is an adapted stochastic process $(f_s, 0 \le s \le T)$ such that

$$(12.1) \qquad F = \mathbb{E}\,(F) + \int_0^T f(s)dB_s \,.$$

This formula has a very broad application to various fields, including finance. For the fractional Brownian motions of parameter $H > 1/2$, an analogous formula is obtained by using the quasi conditional expectation and is applied to finance in [**58**]. However, the uniqueness problem is not discussed there. In this section we study the representation formula for all parameter $H \in (0,1)$ and in particular we shall study the uniqueness problem. Let F be a functional of fractional Brownian motion B_t^H, $0 \le t \le T$. Then by the chaos expansion, we may write

$$(12.2) \qquad F = \mathbb{E}\,(F) + \sum_{n=1}^\infty \frac{1}{n!} \int_{0<t_1,\cdots,t_n<T} f_n(t_1,\cdots,t_n)dB_{t_1}^H \cdots dB_{t_n}^H \,.$$

Assume that $f_n(t_1,\cdots,t_n)$, $n = 1,2,\cdots$, are measurable function of $t_1,\cdots,t_n \in [0,T]$. Define $f_n(t)(t_1,\cdots,t_{n-1}) = f_n(t_1,\cdots,t_{n-1},t)$. If $\sum_{n=1}^\infty \frac{1}{(n-1)!}\|f_n(t)\|_{\Theta_H^{\otimes(n-1)}}^2 < \infty$ for almost all $t \in [0,T]$, then we define the candidate for the Clark derivative as

$$(12.3) \qquad f(t) = \sum_{n=1}^\infty \frac{1}{(n-1)!} \int_{0<t_1,\cdots,t_{n-1}<t} f_n(t_1,\cdots,t_{n-1},t)dB_{t_1}^H \cdots dB_{t_{n-1}}^H \,.$$

Then it is easy to check that $f(t)$ is $\mathcal{F}_t$-adapted and

$$F = \mathbb{E}\,(F) + \int_0^T f(t)dB_t^H \,.$$

Now we consider the uniqueness of $f(t)$. If there is no requirement on $f(t)$, then such $f(t)$ may not be unique. Here is one example. From Example 7.5 we have

$$\left(B_T^H\right)^2 \;=\; \mathbb{E}\,\left(B_T^H\right)^2 + 2\int_0^T B_s^H dB_s^H$$

On the other hand,

$$\int_0^T B_T^H dB_t^H \;=\; I_2(\chi_{[0,T]}^{\otimes 2})$$
$$=\; \left(B_T^H\right)^2 - T^{2H} \,.$$

Therefore

$$(B_T^H)^2 = T^{2H} + \int_0^T B_T^H \, dB_t^H \,.$$

This means that both $2B_s^H, 0 \le s \le T$ and $B_T^H, 0 \le s \le T$ are candidates for the "Clark derivative" of $\left(B_T^H\right)^2$. The main problem is that B_T^H is not adapted to the filtration $(\mathcal{F}_t^H, 0 \le t \le T)$. Although the adaptedness cannot be used to simplify the theory of stochastic analysis of fBm it is useful to determine the uniqueness of the Clark derivative.

THEOREM 12.1. *Let $F \in L^2(\Omega, \mathcal{F}, P^H)$ be given by (12.2) such that $f_n, n = 1, 2, \cdots$ are measurable functions of $t_1, \cdots, t_n \in [0, T]$ and $\displaystyle\sum_{n=1}^{\infty} \frac{1}{(n-1)!} \|f_n(t)\|_{\Theta_H^{\otimes(n-1)}}^2 < \infty$ for almost all $t \in [0, T]$. Then there is a unique $f(t)$ which is adapted to the filtration $(\mathcal{F}_t, 0 \le t \le T)$ such that*

$$(12.4) \qquad\qquad F = \mathbb{E}(F) + \int_0^T f(t) dB_t^H$$

and $f(t)$ is given by (12.3).

Proof   We need to show the uniqueness. It suffices to show that if $\int_0^T f(t) dB_t^H = 0$ and $f(t) \in \mathcal{F}_t$ for all $t \in [0, T]$, then $f(t) = 0$ for all $t \in [0, T]$. Because $f(t) \in \mathcal{F}_t$, there is a chaos expansion for $f(t)$, namely, there are $f_n(t; t_1, \cdots, t_n)$ such that

$$f(t) = f_0 + \sum_{n=1}^{\infty} \frac{1}{n!} I_n^{H,t}(f_n(t)) \,,$$

where $f_n(t; t_1, \cdots, t_n) = 0$ if any of $t_1, \cdots, t_n$ is greater than t. By definition

$$\int_0^T f(t) dB_t^H = f_0 B_T^H + \sum_{n=1}^{\infty} \frac{1}{n!} I_n^{H,T}(h_n) \,,$$

where h_n is the symmetrization of $f_{n-1}(t_n; t_1, \cdots, t_{n-1})$, *i.e.*

$$h_n(t_1, \cdots, t_n) = \frac{1}{n} \sum_{k=0}^{n} f_{n-1}(t_k, t_1; \cdots, \hat{t}_k, \cdots, t_n) \,,$$

where $\hat{t}_k$ means t_k is missing. Let $G = I_n(g_n)$, where g_n is arbitrary but such that $g_n(t_1, \cdots, t_n) = 0$ if any of $t_1, \cdots, t_n$ is greater than t. First we have

$$\begin{aligned}
0 &= \mathbb{E}\left(G \int_0^T f(t) dB_t^H\right) \\
&= \mathbb{E}\left[I_n^{H,T}(h_n) I_n^{H,T}(g_n)\right] \\
&= n! \int_{[0,t]^n} \mathbb{\Gamma}_{H,T}^{\otimes n} h_n(t_1, \cdots, t_n) \mathbb{\Gamma}_{H,T}^{\otimes n} g_n(t_1, \cdots, t_n) dt_1 \cdots, t_n \\
&= n! \int_{[0,t]^n} (\mathbb{\Gamma}_{H,T}^*)^{\otimes n} \mathbb{\Gamma}_{H,T}^{\otimes n} h_n(t_1, \cdots, t_n) g_n(t_1, \cdots, t_n) dt_1 \cdots, t_n \,.
\end{aligned}$$

Since y is arbitrary, we have $(\mathbb{\Gamma}_{H,T}^*)^{\otimes n} \mathbb{\Gamma}_{H,T}^{\otimes n} h_n(t_1, \cdots, t_n) = 0$. Thus we have

$$h_n(t_1, \cdots, t_n) = (\mathbb{B}_{H,T} \mathbb{B}_{H,T}^*)^{\otimes n} (\mathbb{\Gamma}_{H,T}^*)^{\otimes n} \mathbb{\Gamma}_{H,T}^{\otimes n} h_n(t_1, \cdots, t_n) = 0 \,.$$

Thus $h_n(t_1, \cdots, t_{n-1}, t) = 0$ for all $0 \le t_1, \cdots, t_{n-1} \le t$. However, when $0 \le t_1, \cdots, t_{n-1} < t$, $h_n(t_1, \cdots, t_{n-1}, t) = \frac{1}{n} f_{n-1}(t; t_1, \cdots, t_{n-1})$. This implies that $f(t) = 0$ a.s. proving the theorem. $\square$

EXAMPLE 12.2. In [**58**], we obtain a Clark type representation for $f(B_T^H)$ for $H > 1/2$. We shall give a similar formula for all H. We find a simpler way, which is similar to the method of [**49**]. Let $P_t(x,y) = \frac{1}{\sqrt{2\pi t}} e^{-\frac{|x-y|^2}{2}}$ be the usual heat kernel. Denote $P_t f(x) = \int_{\mathbb{R}} P_t(x,y) f(y) dy$. Consider

$$h(t, B_t^H) = P_{T^{2H} - t^{2H}} f(B_t^H).$$

Applying the Itô formula, noticing $h(T, B_T^H) = f(B_T^H)$ and $h(0, B_0^H) = \mathbb{E}(f(B_T^H))$, we obtain

$$
\begin{aligned}
f(B_T^H) &= \mathbb{E}(f(B_T^H)) + \int_0^T \frac{\partial}{\partial x} P_{T^{2H} - t^{2H}} f(B_t^H) dB_t^H \\
&\quad + H \int_0^T \frac{\partial^2}{\partial x^2} P_{T^{2H} - t^{2H}} f(B_t^H) dt - 2H \int_0^T t^{2H-1} \frac{\partial}{\partial t} P_{T^{2H} - t^{2H}} f(B_t^H) dt.
\end{aligned}
$$

Since the heat kernel satisfies $\frac{\partial P_t}{\partial t} = \frac{1}{2} \frac{\partial^2 P_t}{\partial x^2}$, we have that the last two terms cancel. Thus

$$f(B_T^H) = \mathbb{E}(f(B_T^H)) + \int_0^T \frac{\partial}{\partial x} P_{T^{2H} - t^{2H}} f(B_t^H) dB_t^H.$$

Hence the Clark derivative is given by $\frac{\partial}{\partial x} P_{T^{2H} - t^{2H}} f(B_t^H)$.

Let $g \in \Theta_{H,T} \cap L^2([0,T])$ be a given deterministic function and let $\|g\|_{\Theta_{H,t}}$ be continuously differentiable as a function of $t \in [0,T]$. Now we find the Clark derivative of $F(\int_0^T g(s) dB_s^H)$. We shall follows the idea of the above proof.

Let $\eta(T,t)$ be a deterministic function to be determined later. Set

$$h(t) = \left[P_{\eta(T,t)} F\right]\left(\int_0^t g(s) dB_s^H\right).$$

An application of Itô formula yields

$$
\begin{aligned}
h(T) &= h(0) + \int_0^T \frac{\partial}{\partial x} \left[P_{\eta(T,t)} F\right]\left(\int_0^t g(s) dB_s^H\right) g(t) dB_t^H \\
&\quad + \int_0^T \left.\frac{\partial}{\partial u}\right|_{u=\eta(T,t)} \left[P_u F\right]\left(\int_0^t g(s) dB_s^H\right) \frac{d}{dt} \eta(T,t) dt \\
&\quad + \frac{1}{2} \int_0^T \frac{\partial^2}{\partial x^2} \left[P_{\eta(T,t)} F\right]\left(\int_0^t g(s) dB_s^H\right) \frac{d}{dt} \|g\|_{\Theta_{H,t}}^2 dt.
\end{aligned}
$$

The sum of the last terms is zero if

$$\frac{d}{dt} \eta(T,t) = -\frac{d}{dt} \|g\|_{\Theta_{H,t}}^2.$$

Set $\eta(T,T) = 0$. Then

$$\eta(T,t) = \|g\|_{\Theta_{H,T}}^2 - \|g\|_{\Theta_{H,t}}^2$$

and

$$\eta(T,0) = \|g\|_{\Theta_{H,T}}^2.$$

Thus we have

THEOREM 12.3. *Let F be continuously differentiable with bounded derivatives and let $\|g\|_{\Theta_{H,t}}$ be continuously differentiable with respect to $t \in [0,T]$. Then*

$$F\left(\int_0^T g(s)dB_s^H\right) = P_{\|g\|_{\Theta_{H,T}}} F(0)$$

(12.5)
$$+ \int_0^T \frac{\partial}{\partial x}\left[P_{\|g\|_{\Theta_{H,T}}^2 - \|g\|_{\Theta_{H,t}}^2} F\right]\left(\int_0^t g(s)dB_s^H\right) g(t)dB_t^H.$$

It is obvious that this implies Example (12.2) if we take $g = 1$.

CHAPTER 13

Continuation

It is an important question to know if a stochastic process has continuous path. The stochastic integral as a process by allowing the upper limit to vary has been investigated for the sBm and semimartingales. In [2] conditions have also been given such that there is a continuous version for the stochastic process $(\int_0^t (t-s)^{-\alpha} \phi_s dB_s, t \geq 0)$ for adapted process ϕ. In this section we give conditions such that $(X_t := \int_0^t f(s) dB_s^H, 0 \leq t \leq T)$ has a continuous version for general H and f. We follow the idea of [55].

THEOREM 13.1. *Let* $X_t = \int_0^t f(s) dB_s^H$, $0 \leq t \leq T$.
(i) Let $0 < H < 1/2$. *If there exists a* p *with* $\frac{1}{H} < p < \infty$ *such that*

$$
(13.1) \quad \mathbb{E} \sup_{0 \leq u, v \leq T} \left(|f(u)| + |f'(u)| + |D_u^H f(v)| + |D_u^H f'(v)| \right.
$$
$$
\left. + |\frac{d}{du} D_u^H f(v)| + |\frac{\partial^2}{\partial u \partial v} D_u^H f(v)| \right)^p < \infty,
$$

then $(X_t, 0 \leq t \leq T)$ *has a continuous version.*
(ii) Let $1/2 < H < 1$. *If there is a* p *with* $\frac{1}{H - \frac{1}{2}} < p < \infty$ *such that*

$$
(13.2) \quad \mathbb{E} \left\{ \sup_{0 \leq u \leq T} |f(u)|^p + \sup_{0 \leq u, v \leq T} |D_u^H f(v)|^p \right\} << \infty,
$$

then $(X_t, 0 \leq t \leq T)$ *has a continuous version.*

Proof Let $0 < a < b < \infty$ be given and arbitrary. Then from (7.6) it follows that

$$
\mathbb{E}|X_b - X_a|^p = \mathbb{E} \left| \int_0^b f(s) \chi_{[a,b]}(s) dB_s^H \right|^p
$$
$$
\leq C_p \left\{ \mathbb{E} \left(\int_0^b \left| \mathbb{I}_{H,T}^* (f \chi_{[a,b]})(s) \right|^2 ds \right)^{p/2} \right.
$$
$$
\left. + \mathbb{E} \left(\int_0^b \int_0^b \left| \mathbb{I}_{H,b}^*(s) \mathbb{I}_{H,b}^*(t) D_t^H f(s) \chi_{[a,b]}(s) \right|^2 \right)^{p/2} \right\}
$$
$$
(13.3) \quad =: C_p(I_1 + I_2).
$$

First we assume that $0 < H < 1/2$. Let us estimate I_1. By (5.24) we have

$$\mathbb{I}^*_{H,b}(f\chi_{[a,b]})(s) \;=\; \kappa_H s^{\frac{1}{2}-H}\frac{d}{ds}\int_{s\vee a}^b t^{H-\frac{1}{2}}(t-s)^{H-\frac{1}{2}}f(t)dt$$

$$= \begin{cases} \left(\frac{1}{2}-H\right)\kappa_H s^{\frac{1}{2}-H}\int_a^b t^{H-\frac{1}{2}}(t-s)^{H-\frac{3}{2}}f(t)dt & 0 < s < a \\[2mm] \kappa_H s^{\frac{1}{2}-H}\frac{d}{ds}\int_s^b t^{H-\frac{1}{2}}(t-s)^{H-\frac{1}{2}}f(t)dt & a < s < b \end{cases}$$

When $0 < s < a$, we have that for any $0 \le \beta < \frac{3}{2}-H$,

$$\mathbb{I}^*_{H,b}(f\chi_{[a,b]})(s) \;\le\; C_H \sup_{0\le t\le T}|f(t)|s^{\frac{1}{2}-H}\int_a^b t^{H-\frac{1}{2}}(t-s)^{H-\frac{3}{2}}dt$$

$$\le\; C_H \sup_{0\le t\le T}|f(t)|s^{\frac{1}{2}-H}\int_a^b s^{H-\frac{1}{2}}(t-a)^{H-\frac{3}{2}+\beta}(a-s)^{-\beta}dt$$

$$\le\; C_{H,\beta,T} \sup_{0\le t\le T}|f(t)|(b-a)^{H-\frac{1}{2}+\beta}(a-s)^{-\beta}.$$

Therefore

$$I_3 \;:=\; \mathbb{E}\left(\int_0^a \left|\mathbb{I}^*_{H,T}f(s)\chi_{[a,b]}(s)\right|^2 ds\right)^{p/2} \sup_{0\le t\le T}|f(t)|^p$$

$$\le\; C_{H,\beta,T}(b-a)^{(H-\frac{1}{2}+\beta)p}\left(\int_0^a (a-s)^{-2\beta}ds\right)^{p/2}\sup_{0\le t\le T}|f(t)|^p$$

$$=\; C_{H,\beta,T}(b-a)^{(H-\frac{1}{2}+\beta)p}\mathbb{E}\left(\sup_{0\le t\le T}|f(t)|^p\right)a^{\frac{-2\beta+1}{2}p}$$

(13.4) $$\le\; C_{H,\beta,T,p}\mathbb{E}\sup_{0\le t\le T}|f(t)|^p(b-a)^{(H-\frac{1}{2}+\beta)p},$$

where we use the fact that $a^{\frac{-2\beta+1}{2}p} \le C_{T,\beta,p}$ if $0 \le \beta < 1/2$. Therefore for any $\gamma < H$, there is a $\beta \in [0,1/2)$ such that $H-\frac{1}{2}+\beta = \gamma$. Thus

(13.5) $$\mathbb{E}\left(\int_0^a \left|\mathbb{I}^*_{H,T}f(s)\chi_{[a,b]}(s)\right|^2 ds\right)^{p/2} \le C_{H,T,\gamma,p}(b-a)^{\gamma p}.$$

When $a < s < b$, we have

$$\mathbb{I}^*_{H,b}(f\chi_{[a,b]})(s) \;=\; \kappa_H s^{\frac{1}{2}-H}\frac{d}{ds}\int_s^b t^{H-\frac{1}{2}}(t-s)^{H-\frac{1}{2}}f(t)dt$$

$$=\; \kappa_H s^{\frac{1}{2}-H}\frac{d}{ds}\int_0^{b-s}(r+s)^{H-\frac{1}{2}}r^{H-\frac{1}{2}}f(r+s)dr$$

$$=\; \left(H-\frac{1}{2}\right)\kappa_H s^{\frac{1}{2}-H}\int_0^{b-s}(r+s)^{H-\frac{3}{2}}r^{H-\frac{1}{2}}f(r+s)dr$$

$$\quad -\kappa_H s^{\frac{1}{2}-H}b^{H-\frac{1}{2}}(b-s)^{H-\frac{1}{2}}f(b)$$

$$\quad +\kappa_H s^{\frac{1}{2}-H}\int_0^{b-s}(r+s)^{H-\frac{1}{2}}r^{H-\frac{1}{2}}f'(r+s)dr$$

(13.6) $$=:\; I_4(t) + I_5(t) + I_6(t).$$

Now for any $0 < \beta < \frac{1}{2} - H$, we have

$$
\begin{aligned}
\int_a^b |I_4(t)|^2 dt &\leq C_{H,T} \sup_{0 \leq t \leq T} |f(t)|^2 \int_a^b s^{1-2H} \left| \int_0^{b-s} (r+s)^{H-\frac{3}{2}} r^{H-\frac{1}{2}} dr \right|^2 ds \\
&\leq C_{H,T} \sup_{0 \leq t \leq T} |f(t)|^2 \int_a^b s^{1-2H} \left| \int_0^{b-s} s^{-\beta} r^{2H-2+\beta} dr \right|^2 ds \\
&\leq C_{H,T} \sup_{0 \leq t \leq T} |f(t)|^2 \int_a^b s^{1-2H-2\beta} (b-s)^{4H-2+2\beta} ds \\
&\leq C_{H,T} \sup_{0 \leq t \leq T} |f(t)|^2 \int_a^b (b-s)^{4H-2+2\beta} ds \\
&\leq C_{H,T} \sup_{0 \leq t \leq T} |f(t)|^2 (b-a)^{4H-1+2\beta}
\end{aligned}
$$

if $1 - 2H - 2\beta \geq 0$. This inequality is equivalent to $4H - 1 + 2\beta < 2H$. Thus for any $\gamma < H$, one can find $0 < \beta < \frac{1}{2} - H$ such that $4H - 1 + 2\beta > 2\gamma$. Consequently, we obtain that for any $\gamma < H$, there is a $C_{H,T,\gamma,p}$ such that

$$
(13.7) \qquad \mathbb{E} \left(\int_a^b |I_4(t)|^2 dt \right)^{p/2} \leq C_{H,T,\gamma,p} \mathbb{E} \left(\sup_{0 \leq t \leq T} |f(t)|^p \right) (b-a)^{\gamma p}.
$$

If $a < s < b$ and $0 < H < 1/2$, then we have $0 < s^{\frac{1}{2}-H} b^{H-\frac{1}{2}} < 1$. Thus

$$
\begin{aligned}
\int_a^b |I_5(t)|^2 dt &\leq C_H \sup_{0 \leq t \leq T} |f(t)|^2 \int_a^b (b-s)^{2H-1} ds \\
&= C_H (b-a)^{2H} \sup_{0 \leq t \leq T} |f(t)|^2.
\end{aligned}
$$

Therefore

$$
(13.8) \qquad \mathbb{E} \left(\int_a^b |I_5(t)|^2 dt \right)^{p/2} \leq C_{H,T,\gamma,p} \mathbb{E} \left(\sup_{0 \leq t \leq T} |f(t)|^p \right) (b-a)^{Hp}
$$

To estimate $I_6(s)$ we have

$$
\begin{aligned}
|I_6(t)| &\leq C_H s^{\frac{1}{2}-H} \int_0^{b-s} r^{2H-1} dr \sup_{\leq t \leq T} |f'(t)| \\
&= C_H s^{\frac{1}{2}-H} (b-s)^{2H} dr \sup_{\leq t \leq T} |f'(t)|.
\end{aligned}
$$

Hence

$$
(13.9) \quad \mathbb{E} \left(\int_a^b |I_6(t)|^2 dt \right)^{p/2} \leq C_{H,T,p} \mathbb{E} \left(\sup_{\leq t \leq T} |f'(t)|^p \right) (b-a)^{(4H+1)p/2}.
$$

Combining (13.5)-(13.9) we have that for any $\gamma < H$, there is a positive constant $C_{H,T,\gamma,p}$ such that
(13.10)

$$
\mathbb{E} \left\{ \int_a^b \left| \mathbb{T}^*_{H,b}(f\chi_{[a,b]})(s) \right|^2 ds \right\}^{p/2} \leq C_{H,T,\gamma,p} \mathbb{E} \left(\sup_{0 \leq t \leq T} (|f(t)| + |f'(t)|)^p \right) (b-a)^{\gamma p}.
$$

Now we estimate I_2. By the expression for $\mathbb{I}^*_{H,b}$, we have

$$
\mathbb{I}^*_{H,b}(s)\mathbb{I}^*_{H,b}(t)D^H_t f(s)\chi_{[a,b]}(s)
$$

$$
= \kappa^2_H s^{\frac{1}{2}-H}t^{\frac{1}{2}-H}\frac{\partial^2}{\partial s\partial t}\int_s^b du\int_{t\vee a}^b dv u^{H-\frac{1}{2}}v^{H-\frac{1}{2}}(u-s)^{H-\frac{1}{2}}(v-t)^{H-\frac{1}{2}}D^H_u f(v)
$$

$$
= \begin{cases} \kappa^2_H s^{\frac{1}{2}-H}t^{\frac{1}{2}-H}\frac{\partial^2}{\partial s\partial t}\int_s^b du\int_a^b dv u^{H-\frac{1}{2}}v^{H-\frac{1}{2}}(u-s)^{H-\frac{1}{2}}(v-t)^{H-\frac{1}{2}}D^H_u f(v) \\ \qquad\qquad \text{when }\; 0<t<a \\[6pt] \kappa^2_H s^{\frac{1}{2}-H}t^{\frac{1}{2}-H}\frac{\partial^2}{\partial s\partial t}\int_s^b du\int_t^b dv u^{H-\frac{1}{2}}v^{H-\frac{1}{2}}(u-s)^{H-\frac{1}{2}}(v-t)^{H-\frac{1}{2}}D^H_u f(v) \\ \qquad\qquad \text{when }\; a<t<b \end{cases}
$$

If $0<t<a$, then

$$
\mathbb{I}^*_{H,b}(s)\mathbb{I}^*_{H,b}(t)D^H_t f(s)\chi_{[a,b]}(s)
$$

$$
= \left(\frac{1}{2}-H\right)\kappa^2_H s^{\frac{1}{2}-H}t^{\frac{1}{2}-H}\frac{\partial}{\partial s}\int_0^{b-s} dr\int_a^b dv(r+s)^{H-\frac{1}{2}}v^{H-\frac{1}{2}}r^{H-\frac{1}{2}}(v-t)^{H-\frac{3}{2}}D^H_{r+s}f(v)
$$

$$
= -\left(\frac{1}{2}-H\right)\kappa^2_H s^{\frac{1}{2}-H}t^{\frac{1}{2}-H}b^{H-\frac{1}{2}}(b-s)^{H-\frac{1}{2}}\int_a^b v^{H-\frac{1}{2}}(v-t)^{H-\frac{3}{2}}D^H_b f(v)dv
$$

$$
-\left(\frac{1}{2}-H\right)^2\kappa^2_H s^{\frac{1}{2}-H}t^{\frac{1}{2}-H}\int_0^{b-s} dr\int_a^b (r+s)^{H-\frac{3}{2}}
$$

$$
v^{H-\frac{1}{2}}r^{H-\frac{1}{2}}(v-t)^{H-\frac{3}{2}}D^H_{r+s}f(v)dv
$$

$$
+\left(\frac{1}{2}-H\right)\kappa^2_H s^{\frac{1}{2}-H}t^{\frac{1}{2}-H}\int_0^{b-s} dr\int_a^b (r+s)^{H-\frac{1}{2}}
$$

$$
v^{H-\frac{1}{2}}r^{H-\frac{1}{2}}(v-t)^{H-\frac{3}{2}}\frac{d}{ds}D^H_{r+s}f(v)dv\,.
$$

Therefore

$$
|\mathbb{I}^*_{H,b}(s)\mathbb{I}^*_{H,b}(t)D^H_t f(s)\chi_{[a,b]}(s)|
$$

$$
= C_H \sup_{0\leq u,v\leq T}|D^H_u f(v)|s^{\frac{1}{2}-H}t^{\frac{1}{2}-H}\left\{b^{H-\frac{1}{2}}(b-s)^{H-\frac{1}{2}}\int_a^b v^{H-\frac{1}{2}}(v-t)^{H-\frac{3}{2}}dv\right.
$$

$$
+\int_0^{b-s}\int_a^b (r+s)^{H-\frac{3}{2}}v^{H-\frac{1}{2}}r^{H-\frac{1}{2}}(v-t)^{H-\frac{3}{2}}dv
$$

$$
\left.+C_H \sup_{0\leq u,v\leq T}|\frac{d}{du}D^H_u f(v)|s^{\frac{1}{2}-H}t^{\frac{1}{2}-H}\int_0^{b-s}\int_a^b (r+s)^{H-\frac{1}{2}}v^{H-\frac{1}{2}}r^{H-\frac{1}{2}}(v-t)^{H-\frac{3}{2}}dv\right\}
$$

$$
=: I_7 + I_8 + I_9\,.
$$

We estimate I_7, I_8, and I_9 separately. First we estimate I_7. Let β be any number such that $\frac{1}{2} - H < \beta < 1/2$.

$$
\begin{aligned}
I_7 &\leq C_{H,T} \sup_{0 \leq u,v \leq T} |D_u^H f(v)| t^{\frac{1}{2}-H} (b-s)^{H-\frac{1}{2}} \int_a^b v^{H-\frac{1}{2}} (v-t)^{H-\frac{3}{2}} dv \\
&\leq C_{H,T} \sup_{0 \leq u,v \leq T} |D_u^H f(v)| t^{\frac{1}{2}-H} (b-s)^{H-\frac{1}{2}} a^{H-\frac{1}{2}} \int_a^b (v-t)^{H-\frac{3}{2}} dv \\
&\leq C_{H,T} \sup_{0 \leq u,v \leq T} |D_u^H f(v)| (b-s)^{H-\frac{1}{2}} \int_a^b (v-a)^{H-\frac{3}{2}+\beta} (a-t)^{-\beta} dv \\
&= C_{H,T} \sup_{0 \leq u,v \leq T} |D_u^H f(v)| (b-s)^{H-\frac{1}{2}} (b-a)^{H-\frac{1}{2}+\beta} (a-t)^{-\beta}.
\end{aligned}
$$

Therefore for any $\gamma < H$, there is a positive constant $C_{H,T,p,\gamma}$ such that

$$
\begin{aligned}
\mathbb{E} \left(\int_0^b \int_0^a I_7^2 ds dt \right)^{p/2} &\leq C_{H,T,p,\gamma} \mathbb{E} \sup_{0 \leq u,v \leq T} |D_u^H f(v)|^p (b-a)^{(H-\frac{1}{2}+\beta)p} \\
&\leq C_{H,T,p,\gamma} \mathbb{E} \sup_{0 \leq u,v \leq T} |D_u^H f(v)|^p (b-a)^{\gamma p}.
\end{aligned}
$$
(13.11)

Now we estimate I_8. We have for any $0 < \beta < 1/2$

$$
\begin{aligned}
I_8 &\leq C_H \sup_{0 \leq u,v \leq T} |D_u^H f(v)| s^{\frac{1}{2}-H} \int_0^{b-s} (r+s)^{H-\frac{3}{2}} r^{H-\frac{1}{2}} dr\, t^{\frac{1}{2}-H} \int_a^b v^{H-\frac{1}{2}} (v-t)^{H-\frac{3}{2}} dv \\
&\leq C_{H,T} \sup_{0 \leq u,v \leq T} |D_u^H f(v)| s^{-H} \int_0^{b-s} r^{2H-\frac{3}{2}} dr \int_a^b (v-t)^{H-\frac{3}{2}} dv \\
&= C_{H,T} \sup_{0 \leq u,v \leq T} |D_u^H f(v)| s^{-H} (b-s)^{2H-\frac{1}{2}} (b-a)^{H-\frac{1}{2}+\beta} (a-t)^{-\beta}.
\end{aligned}
$$

Therefore for any $0 < \gamma < H$

$$
\mathbb{E} \left(\int_0^b \int_0^a I_8^2 ds dt \right)^{p/2} \leq C_{H,T,p,\gamma} \mathbb{E} \sup_{0 \leq u,v \leq T} |D_u^H f(v)|^p (b-a)^{\gamma p}.
$$
(13.12)

In a similar way we have

$$
\mathbb{E} \left(\int_0^b \int_0^a I_9^2 ds dt \right)^{p/2} \leq C_{H,T,p,\gamma} \mathbb{E} \sup_{0 \leq u,v \leq T} |\frac{d}{du} D_u^H f(v)|^p (b-a)^{\gamma p}.
$$
(13.13)

If $a < t < b$, then

$$\mathbb{I}^*_{H,b}(s)\mathbb{I}^*_{H,b}(t)D^H_t f(s)\chi_{[a,b]}(s)$$

$$= \kappa^2_H s^{\frac{1}{2}-H} t^{\frac{1}{2}-H} \frac{\partial^2}{\partial s \partial t} \int_s^b du \int_t^b dv u^{H-\frac{1}{2}} v^{H-\frac{1}{2}} (u-s)^{H-\frac{1}{2}} (v-t)^{H-\frac{3}{2}} D^H_{r+s} f(v)$$

$$= \kappa^2_H s^{\frac{1}{2}-H} t^{\frac{1}{2}-H} \frac{\partial^2}{\partial s \partial t} \int_0^{b-s} d\xi \int_0^{b-t} d\eta (\xi+s)^{H-\frac{1}{2}} (\eta+t)^{H-\frac{1}{2}} \xi^{H-\frac{1}{2}} \eta^{H-\frac{3}{2}} D^H_{\xi+s} f(\eta+t)$$

$$= \kappa^2_H s^{\frac{1}{2}-H} t^{\frac{1}{2}-H} \Bigg\{ b^{H-\frac{1}{2}} (b-s)^{H-\frac{1}{2}} b^{H-\frac{1}{2}} (b-t)^{H-\frac{1}{2}} D^H_b f(b)$$

$$+ (\frac{1}{2}-H) \int_0^{b-s} (\xi+s)^{H-\frac{3}{2}} \xi^{H-\frac{1}{2}} b^{H-\frac{1}{2}} (b-t)^{H-\frac{1}{2}} D^H_{\xi+s} f(b) d\xi$$

$$+ \int_0^{b-s} (\xi+s)^{H-\frac{1}{2}} \xi^{H-\frac{1}{2}} b^{H-\frac{1}{2}} (b-t)^{H-\frac{1}{2}} \frac{d}{ds} D^H_{\xi+s} f(b) d\xi$$

$$+ \left(\frac{1}{2}-H\right) \int_0^{b-t} b^{H-\frac{1}{2}} (b-s)^{H-\frac{1}{2}} (\eta+t)^{H-\frac{3}{2}} \eta^{H-\frac{1}{2}} D^H_b f(\eta+t) d\eta$$

$$+ \left(\frac{1}{2}-H\right)^2 \int_0^{b-t} d\xi \int_0^{b-s} d\eta (\xi+s)^{H-\frac{3}{2}} \xi^{H-\frac{1}{2}} (\eta+t)^{H-\frac{3}{2}} \eta^{H-\frac{1}{2}} D^H_{\xi+s} f(\eta+t)$$

$$+ \left(H-\frac{1}{2}\right) \int_0^{b-t} d\xi \int_0^{b-s} d\eta (\xi+s)^{H-\frac{1}{2}} \xi^{H-\frac{1}{2}} (\eta+t)^{H-\frac{3}{2}} \eta^{H-\frac{1}{2}} \frac{d}{ds} D^H_{\xi+s} f(\eta+t) d\eta$$

$$+ \int_0^{b-t} b^{H-\frac{1}{2}} (b-s)^{H-\frac{1}{2}} (\eta+t)^{H-\frac{1}{2}} \eta^{H-\frac{1}{2}} \frac{d}{dt} D^H_b f(\eta+t) d\eta$$

$$+ \left(H-\frac{1}{2}\right) \int_0^{b-t} d\xi \int_0^{b-s} d\eta (\xi+s)^{H-\frac{3}{2}} \xi^{H-\frac{1}{2}} (\eta+t)^{H-\frac{1}{2}} \eta^{H-\frac{1}{2}} \frac{d}{ds} D^H_{\xi+s} f(\eta+t)$$

$$+ \int_0^{b-t} d\xi \int_0^{b-s} d\eta (\xi+s)^{H-\frac{1}{2}} \xi^{H-\frac{1}{2}} (\eta+t)^{H-\frac{1}{2}} \eta^{H-\frac{1}{2}} \frac{\partial^2}{\partial s \partial t} D^H_{\xi+s} f(\eta+t)$$

$$=: I_{10} + I_{11} + I_{12} + I_{13} + I_{14} + I_{15} + I_{16} + I_{17} + I_{18} \,.$$

We will show for all of these terms that for any $\gamma < H$, there is a positive constant $C_{H,T,P,\gamma}$ such that

$$\mathbb{E} \left(\int_0^b \int_a^b I^2_k ds dt \right)^{p/2} \leq C_{H,T,P} \mathbb{E} \sup_{0 \leq u, v \leq T} \left(|D^H_u f(v)|^p + |\frac{d}{du} D^H_u f(v)|^p \right.$$

$$(13.14) \qquad\qquad \left. + |\frac{d}{dv} D^H_u f(v)|^p + |\frac{\partial^2}{\partial u \partial v} D^H_u f(v)|^p \right) (b-a)^{\gamma p} \,,$$

where $k = 10, 11, \cdots, 18$. Let us consider for example I_{14}. Let $0 < \beta < \frac{1}{2} - H$. For any $0 < \gamma < H$ we have

$$I_{14}^2 \leq C_H \int_0^b \int_a^b \left[(st)^{\frac{1}{2}-H} \int_0^{b-s} \int_0^{b-t} (\xi+s)^{H-\frac{3}{2}} (\eta+t)^{H-\frac{3}{2}} \right.$$

$$\left. (\xi\eta)^{H-\frac{1}{2}} |D_{\xi+s}^H f(\eta+t)| d\xi d\eta \right]^2 ds dt$$

$$\leq C_H \int_0^b \int_a^b \left[(st)^{1-2H}(st)^{-2\beta} \left(\int_0^{b-s} \xi^{2H-2+\beta} d\xi \int_0^{b-t} \eta^{2H-2+\beta} d\eta \right)^2 \right]$$

$$\leq C_{H,T}(b-a)^{4H+2\beta-1} \sup_{0 \leq u,v \leq T} |D_u^H f(v)|^2$$

$$\leq C_{H,T,\gamma}(b-a)^{2\gamma}.$$

Therefore (13.14) is true for $k = 14$. Other cases can be treated in similar way.

Thus we have showed that for any $\gamma < H$ there is constant $C_{H,T,p,\gamma}$ such that

$$I_2 \leq C_{H,T,p,\gamma} \mathbb{E} \sup_{0 \leq u,v \leq T} \left(|D_u^H f(v)| + |D_u^H f'(v)| + |\frac{d}{du}D_u^H f(v)| + |\frac{d}{dv}D_u^H f(v)| \right.$$

$$(13.15) \qquad \left. + |\frac{\partial^2}{\partial u \partial v}D_u^H f(v)| \right)^p (b-a)^{p\gamma}.$$

By Kolmogorov theorem (13.3), (13.10) and (13.15) imply Part (i) of the theorem.

Now we assume that $1/2 < H < 1$. From (5.21) it follows that

$$\mathbb{I}_{H,T}^* \left(f\chi_{[a,b]} \right)(s) = \left(H - \frac{1}{2} \right) \kappa_H s^{\frac{1}{2}-H} \int_{s\vee a}^b t^{H-\frac{1}{2}} (t-s)^{H-\frac{3}{2}} f(t)dt.$$

Thus when $a \leq s \leq b$

$$|\mathbb{I}_{H,T}^* \left(f\chi_{[a,b]} \right)(s)| \leq C_{H,T} s^{\frac{1}{2}-H} (b-s)^{H-\frac{1}{2}} \sup_{0 \leq t \leq T} |f(t)|.$$

Let $r_1, r_2 \geq 0$, $\frac{1}{r_1} + \frac{1}{r_2} = 1$, and $(1-2H)r_1 > -1$. Then

$$\int_a^b |\mathbb{I}_{H,T}^* \left(f\chi_{[a,b]} \right)(s)|^2 ds$$

$$\leq C_{H,T} \sup_{0 \leq t \leq T} |f(t)|^2 \int_a^b s^{1-2H}(b-s)^{2H-1} ds$$

$$\leq C_{H,T} \left(\int_a^b s^{(1-2H)r_1} ds \right)^{1/r_1} \left(\int_a^b s^{(2H-1)r_1} ds \right)^{1/r_2} \sup_{0 \leq t \leq T} |f(t)|^2$$

$$\leq C_{H,T,r_1}(b-a)^{2H-\frac{1}{r_1}} \sup_{0 \leq t \leq T} |f(t)|^2$$

$$\leq C_{H,T,r_1}(b-a)^\rho \sup_{0 \leq t \leq T} |f(t)|^2$$

for any $0 < \rho < 1$.

If $0 < s < a$, then

$$\mathbb{I}_{H,T} \left(f\chi_{[a,b]} \right)(s) \leq C_{H,T,p} s^{\frac{1}{2}-H}(b-a)^{H-\frac{1}{2}} \sup_{0 \leq t \leq T} |f(t)|.$$

Therefore when $0 < s < a$, we have

$$\left| \mathbb{I}^*_{H,T} \left(f\chi_{[a,b]} \right) (s) \right| \leq C_{H,T} s^{\frac{1}{2}-H} (b-a)^{H-\frac{1}{2}} \sup_{0 \leq t \leq T} |f(t)|.$$

Hence,

$$(13.16) \quad \mathbb{E} \left[\int_0^b \left| \mathbb{I}^*_{H,T} \left(f\chi_{[a,b]} \right) (s) \right|^2 ds \right]^{p/2} \leq C_{H,T,p} \mathbb{E} \sup_{0 \leq t \leq T} |f(t)|^p (b-a)^{(H-\frac{1}{2})p}.$$

This is an estimate for I_1 defined by (13.3) in the case $1/2 < H < 1$. Now we are going to estimate I_2 defined by (13.3)

$$\mathbb{I}^*_{H,b}(s) \mathbb{I}^*_{H,T}(t) D_t^H \left(f(s)\chi_{[a,b]}(s) \right)$$

$$= \left(H - \frac{1}{2} \right)^2 \kappa_H^2 s^{\frac{1}{2}-H} t^{\frac{1}{2}-H} \int_0^b du \int_{a \vee t}^b dv\, u^{H-\frac{1}{2}} v^{H-\frac{1}{2}} (u-s)^{H-\frac{3}{2}} (v-t)^{H-\frac{3}{2}} D_u^H f(v).$$

$$\left| \mathbb{I}^*_{H,b}(s) \mathbb{I}^*_{H,T}(t) D_t^H \left(f(s)\chi_{[a,b]} \right) \right|$$

$$\leq C_H s^{\frac{1}{2}-H} t^{\frac{1}{2}-H} \int_0^b du \int_{a \vee t}^b dv\, u^{H-\frac{1}{2}} v^{H-\frac{1}{2}} (u-s)^{H-\frac{3}{2}} (v-t)^{H-\frac{3}{2}} \sup_{0 \leq u,v \leq T} |D_u^H f(v)|$$

$$\leq C_{H,T} s^{\frac{1}{2}-H} t^{\frac{1}{2}-H} \int_{a \vee t}^b (v-t)^{H-\frac{3}{2}} dv \sup_{0 \leq u,v \leq T} |D_u^H f(v)|$$

$$\leq C_{H,T} s^{\frac{1}{2}-H} t^{\frac{1}{2}-H} \sup_{0 \leq u,v \leq T} |D_u^H f(v)| (b-a \vee t)^{H-\frac{1}{2}}.$$

Therefore

$$(13.17) \quad \mathbb{E} \left(\int_0^b \int_0^b \left| \mathbb{I}^*_{H,b}(s) \mathbb{I}^*_{H,T}(t) D_t^H \left(f(s)\chi_{[a,b]} \right) \right|^2 ds dt \right)^{p/2}$$

$$\leq C_{H,T,p} \left(\int_0^b \int_0^b s^{1-2H} t^{1-2H} (b-a)^{2H-1} ds dt \right)^{p/2} \mathbb{E} \left(\sup_{0 \leq u,v \leq T} |D_u^H f(v)|^p \right)$$

$$\leq C_{H,T,p} \mathbb{E} \left(\sup_{0 \leq u,v \leq T} |D_u^H f(v)|^p \right) (b-a)^{(H-\frac{1}{2})p}.$$

Therefore
(13.18)

$$\mathbb{E} |X_b - X_a|^p \leq C_{H,T} \mathbb{E} \left(\sup_{0 \leq t \leq T} |f(t)|^p + \sup_{0 \leq u,v \leq T} |D_u^H f(v)|^p \right) (b-a)^{(H-\frac{1}{2})p}.$$

By Kolmogorov theorem $(X_t, 0 \leq t \leq T)$ has continuous modification. $\qquad \square$

CHAPTER 14

Stochastic Control

Optimal control of a (controllable) stochastic differential system has been studied since long time. Numerous results and applications have been found. Many books have been written. There are many effective approaches to solve the problem of optimal control of a controlled stochastic system. Among them two are very famous. One approach is the so-called *dynamic programming*. This approach reduces the stochastic optimal control problem (of a Markov system) to a *Hamilton-Jacobi-Bellman equation*, usually a nonlinear partial differential equation which has been studied extensively (see [12], [39] and the references therein). Another method is the so-called *Pontryagin's maximum principle* (which involves Hamiltonian systems). The most important and elementary stochastic optimal control problem is the linear quadratic problem (see [113]), where the controlled system is determined by a linear stochastic differential equation and the performance functional is quadratic with respect to the control and the state. In this case both approaches mentioned above lead to explicit solution for the optimal control as well as the optimal value function by first solving a Riccati type equation.

In a preprint [62] the linear quadratic problem is considered for fractional Brownian motion of Hurst parameter $H > 1/2$. We shall study the linear quadratic optimal control problem for a system driven by fractional Brownian motion of any Hurst parameter.

Let $0 < H < 1$ and let a_t, b_t be bounded measurable functions of $t \in [0,T]$. As before let $(B_t^H, 0 \leq t \leq T)$ be a fractional Brownian motion of Hurst parameter H. The linear stochastic differential system (geometric fractional Brownian motion) is given by the following (one-dimensional) equation

$$(14.1) \qquad \begin{cases} dx_t = a_t x_t dt + b_t x_t dB_t^H, \quad 0 \leq t \leq T \\ x_0 = x \end{cases}$$

If $H > 1/2$ the existence and uniqueness of the solution to this equation are known in [58], [13]. For general Hurst parameter $H \in (0,1)$ we have the following theorem concerning the solvability.

THEOREM 14.1. *Let $0 < H < 1$. Let a_t and b_t be continuous functions of $t \in [0,T]$. If $0 < H < 1/2$, then we assume that $(b_t, 0 \leq t \leq T)$ is continuously differentiable. Then Equation (14.1) has a unique solution, which is given by*

$$(14.2) \qquad x_t = x \exp \left\{ \int_0^t b_s dB_s^H - \frac{1}{2} \|b\|_{\Theta_{H,t}}^2 + \int_0^t a_s ds \right\}.$$

Proof First let us prove that x_t defined by (14.2) is the solution to (14.1). Applying the Itô formula (Theorem 11.1), we have

$$dx_t = x_t \left[b_t dB_t^H - \frac{1}{2} \frac{d}{dt} \|b\|_{\Theta_{H,t}}^2 dt + a_t dt \right] + \frac{1}{2} x_t \frac{d}{dt} \|b\|_{\Theta_{H,t}}^2 dt$$

$$= x_t \left[b_t dB_t^H + a_t dt \right] .$$

Thus x_t satisfies Equation (14.1). To show the uniqueness we define $\xi_t = e^{-\int_0^t a_s ds}$. Then $y_t := \xi_t x_t$ satisfies

$$dy_t = y_t b_t dB_t^H .$$

Thus by iteration we have

$$y_t = x + \int_0^t y_s b_s dB_s^H$$

$$= x + \int_0^t b_s x dB_s^H + \int_0^t \left(\int_0^s b_u y_u dB_u^H \right) b_s dB_s^H .$$

By iteration we have

$$y_t = x \left(\sum_{n=0}^{\infty} \int_{0<s_1<\cdots<s_n} b_{s_1} \cdots b_{s_n} dB_{s_1}^H \cdots dB_{s_n}^H \right) .$$

Namely, if $y_t = \xi_t x_t$, then it is given by the above series. This proves the uniqueness. □

Formally let x_t be the solution to (14.1). Then

(14.3) $$\mathbb{D}_t^H x_t = r_t x_t ,$$

where

(14.4)
$$\begin{aligned} r_t &= \mathbb{I}_{H,t} \mathbb{I}_{H,t}^* b(t) \\ &= \begin{cases} H(2H-1) \int_0^t |t-s|^{2H-2} b_s ds & 1/2 < H < 1 \\ H t^{2H-1} b_0 + H \int_0^t |t-s|^{2H-1} \text{sign}\,(t-s) b_s' ds & 0 < H < 1/2 \end{cases} \end{aligned}$$

We shall study the following stochastic linear quadratic control problem. The controlled system is given by

(14.5)
$$\begin{cases} dx_t = (a_t x_t + b_t u_t) dt + (c_t x_t + d_t u_t) dB_t^H \\ x_0 = x , \end{cases}$$

where a_t, c_t are real valued continuous functions of t and b_t and d_t are row vector (of dimension m) valued function of t and u_t is the control (column vector function of dimension m). Let Q_t, R_t be continuous function of $t \in [0, T]$ and $H > 0$. Introduce the utility functional of x and u:

(14.6) $$J(x, u) := \mathbb{E} \left[\int_0^T \left(Q_t x_t^2 + u_t^T R_t u_t \right) dt + G x_T^2 \right] .$$

The problem that we are to study is the following
Problem: *Minimize $J(x, u)$ defined by (14.6) subject to (14.5).*

In fact we shall only be interested in the feedback Markov control of the following form $u_t = K_t x_t$, where K_t is an m dimensional column vector valued function of t (it is called feedback gain).

THEOREM 14.2. *Assume that for a.e. $t \in [0, T]$, $d_t = 0$, $Q_t \geq 0$, and $R_t \succ \delta I$ for some given $\delta > 0$, and $G \geq 0$. Let $\|c\|_{\Theta_{H,t}}$ be continuously differentiable of $t \in [0, T]$. Then the following Riccati equation*

(14.7)
$$\begin{cases} \dot{p}_t + 2p_t a_t + p_t \frac{d}{dt} \|c\|^2_{\Theta_{H,t}} + Q_t - b_t R_t^{-1} b_t^* p_t^2 = 0 \\ \\ p_T = G \end{cases}$$

admits a unique solution p over $[0, T]$ with $p_t \geq 0$, $\forall t \in [0, T]$. Moreover, the optimal Markov linear feedback control for the problem (14.5)-(14.6) is given by

(14.8)
$$\hat{u}_t = \hat{K}_t x_t, \quad \text{with} \quad \hat{K}_t = -R_t^{-1} b_t^* p_t.$$

Finally, the optimal value of (14.5)-(14.6) is $p_0 x_0^2$.

Proof The solvability of the (classical) Riccati equation (14.7) was proved in, e.g., [113, p. 297, Corollary 2.10]. Let $u_t = K_t x_t$. Then (14.5) becomes

$$dx_t = (a_t + b_t K_t) x_t dt + c_t x_d dB_t^H.$$

By Theorem 14.1 ((14.2)) we have

$$x_t = x \exp\left\{ \int_0^t c_s dB_s^H - \frac{1}{2} \|c\|^2_{\Theta_{H,t}} + \int_0^t (a_s + b_s K_s) ds \right\}.$$

Therefore

$$x_t^2 = x^2 \exp\left\{ 2 \int_0^t c_s dB_s^H - \|c\|^2_{\Theta_{H,t}} + 2 \int_0^t (a_s + b_s K_s) ds \right\}.$$

By Itô formula (11.2) we have

$$d(x_t^2) = 2x_t^2 \left[a_t + b_t K_t + \frac{1}{2} \frac{d}{dt} \|c\|^2_{\Theta_{H,t}} \right] dt + 2x_t^2 c_t dB_t^H.$$

Thus an integration by parts formula yields

$$d(p_t x_t^2) = x_t^2 \left[\dot{p}_t + 2p_t(a_t + b_t K_t) + p_t \frac{d}{dt} \|c\|^2_{\Theta_{H,t}} \right] dt + 2x_t^2 c_t p_t dB_t^H.$$

Taking integration from 0 to T we get

$$p_T x_T^2 = p_0 x_0^2 + \int_0^T x_t^2 \left[\dot{p}_t + 2p_t(a_t + b_t K_t) + p_t \frac{d}{dt} \|c\|^2_{\Theta_{H,t}} \right] dt + 2 \int_0^T x_t^2 c_t p_t dB_t^H.$$

Denote $f_t = x_t^2 c_t p_t$. Hence

$$\mathbb{E}\left[p_T x_T^2 \right] = p_0 x_0^2 + \mathbb{E} \int_0^T x_t^2 \left[\dot{p}_t + 2p_t(a_t + b_t K_t) + p_t \frac{d}{dt} \|c\|^2_{\Theta_{H,t}} \right] dt.$$

Since $p_T = G$, we obtain

$$
\begin{aligned}
J(x_0, K.) &= p_0 x_0^2 + \mathbb{E} \int_0^T x_t^2 \bigg[\dot{p}_t + 2p_t(a_t + b_t K_t) \\
&\quad + (Q_t + K_t^* R_t K_t) + p_t \frac{d}{dt} \|c\|_{\Theta_{H,t}}^2 \bigg] dt \\
&= p_0 x_0^2 + \mathbb{E} \int_0^T x_t^2 \bigg[\dot{p}_t + 2p_t a_t + p_t \frac{d}{dt} \|c\|_{\Theta_{H,t}}^2 + Q_t \\
&\quad + (K_t + R_t^{-1} b_t^* p_t)^* R_t (K_t + R_t^{-1} b_t^* p_t) - b_t R_t^{-1} b_t^* p_t^2 \bigg] dt \bigg\}
\end{aligned}
$$

$$
(14.9) \qquad = p_0 x_0^2 + \mathbb{E} \int_0^T (K_t + R_t^{-1} b_t^* p_t)^* R_t (K_t + R_t^{-1} b_t^* p_t) dt \,,
$$

where the last equality was due to the Riccati equation (14.7). Eq. (14.9) shows that the cost function achieves its minimum when $\hat{K}_t = -R_t^{-1} b_t^* p_t$, with the minimum value being $p_0 x_0^2$. This proves the theorem. $\qquad\qquad\square$

REMARK 14.3. It is interesting to note that the Riccati equation (14.7) corresponds to the following linear–quadratic control problem with (normal) Brownian motion:

$$(14.10) \qquad \begin{array}{l} \text{Minimize} \quad (14.6) \\[1ex] \text{subject to} \quad \begin{cases} dx_t = (a_t x_t + b_t u_t)dt + \tilde{c}_t x_t dB_t \\[1ex] x_0 \in \mathbb{R} \quad \text{be given and deterministic,} \end{cases} \end{array}$$

where $\tilde{c}_t = \sqrt{\frac{d}{dt}\|c\|_{\Theta_{H,t}}^2}$ if $\frac{d}{dt}\|c\|_{\Theta_{H,t}} \geq 0$ for all $t \geq 0$. This suggests that, in the current setting, the linear–quadratic control problem with fractional Brownian motion is equivalent (in the sense of sharing the same optimal feedback control and optimal value) to a linear–quadratic control problem with Brownian motion where the diffusion coefficient of the state is properly modified.

CHAPTER 15

Appendix

LEMMA 15.1. *For $\mu > 0$, $\nu > 0$, and $c > 1$, we have*

$$(15.1) \qquad \int_0^1 x^{\mu-1}(1-x)^{\nu-1}(\rho-x)^{-\mu-\nu}dx = B(\mu,\nu)\rho^{-\nu}(\rho-1)^{-\mu}.$$

Proof This identity appeared in [**90**]. In fact it is a special case of ([**68**], p.336, formula 3.198) (when $a = 0$, $b = 1$, $c = \rho - 1$). $\square$

LEMMA 15.2. *Let $\mu, \nu > 0$ and $c > 1$. Then*

$$\int_0^1 x^{\mu-1}(1-x)^{\nu-1}(c-x)^{-\mu-\nu+1}dx$$

$$(15.2) \qquad = (\mu+\nu-1)B(\mu,\nu)c^{-\nu+1}\int_0^1 y^{\mu+\nu-2}(c-y)^{-\mu}dy.$$

Proof This is a direct consequence of Lemma 15.1 (see also ([**90**], Lemma 2.2.)) $\square$

LEMMA 15.3. *Let $H > 1/2$, $u > 0$, and $v > 0$. Denote $u \wedge v = \min(u,v)$. Then*

$$\int_0^{u \wedge v} s^{1-2H}(u-s)^{H-\frac{3}{2}}(v-s)^{H-\frac{3}{2}}ds$$

$$(15.3) \qquad = B\left(2-2H, H-\frac{1}{2}\right)u^{\frac{1}{2}-H}v^{\frac{1}{2}-H}|u-v|^{2H-2}.$$

Proof Without loss of generality let us assume $v \le u$. Making substitution $s = vx$, we obtain that the left hand side of (15.3)

$$\int_0^v s^{1-2H}(u-s)^{H-\frac{3}{2}}(v-s)^{H-\frac{3}{2}}ds = v^{-1}\int_0^1 x^{1-2H}\left(\frac{u}{v}-x\right)^{H-\frac{3}{2}}(1-x)^{H-\frac{3}{2}}dx.$$

In Lemma 15.1 taking $\mu = 2 - 2H$, $\nu = H - \frac{1}{2}$, and $\rho = u/v$, we have

$$v^{-1}\int_0^1 x^{1-2H}\left(\frac{u}{v}-x\right)^{H-\frac{3}{2}}(1-x)^{H-\frac{3}{2}}dx$$

$$= B\left(2-2H, H-\frac{1}{2}\right)u^{\frac{1}{2}-H}v^{\frac{1}{2}-H}|u-v|^{2H-2}.$$

This proves the lemma. $\square$

LEMMA 15.4. *Let $\mu, \nu > 0$, and $0 < c < d < \infty$. Then*

$$(15.4) \qquad \int_c^d x^{-\mu-\nu}(x-c)^{\mu-1}(d-x)^{\nu-1}dx = B(\mu,\nu)(d-c)^{\mu+\nu-1}c^{-\nu}d^{-\mu}.$$

Proof Make substitution $x = c + (d-c)\xi$. Then

$$\int_c^d x^{-\mu-\nu}(x-c)^{\mu-1}(d-x)^{\nu-1}dx$$

$$= (d-c)^{\mu+\nu-1}c^{-\mu-\nu}\int_0^1 \xi^{\mu-1}(1-\xi)^{\nu-1}(1+\frac{d-c}{c}\xi)^{-\mu-\nu}d\xi$$

$$= B(\mu,\nu)(d-c)^{\mu+\nu-1}c^{-\mu-\nu}\left(1+\frac{d-c}{c}\right)^{-\mu}$$

The last equality follows from ([**68**], p. 335, Formula 3.197, 4). A simplification yields (15.4). $\square$

Acknowledgement: The author thanks the referee for the critical and constructive comments.

Bibliography

[1] Alòs, E.; León, J. A. and Nualart, D. Stochastic Stratonovich calculus fBm for fractional Brownian motion with Hurst parameter less than 1/2. Taiwanese J. Math. 5 (2001), no. 3, 609-632.

[2] Alòs, E.; Mazet, O. and Nualart, D. Stochastic calculus with respect to fractional Brownian motion with Hurst parameter lesser than $\frac{1}{2}$. Stochastic Process. Appl. 86 (2000), 121-139.

[3] Alòs, E.; Mazet, O. and Nualart, D. Stochastic calculus with respect to Gaussian processes. Ann. Probab. 29 (2001), 766-801.

[4] Bender, C. An Itô formula for generalized functionals of a fractional Brownian motion with arbitrary Hurst parameter. Stochastic Process. Appl. 104 (2003), no. 1, 81–106.

[5] Bender, C. The fractional Itô integral, change of measure and absence of arbitrage. To appear in Proc. of the Royal Soc.

[6] Beran, J. Statistics for long-memory processes. Monographs on Statistics and Applied Probability, 61. Chapman and Hall, New York, 1994.

[7] Berezin, F. A. The method of second quantization. Translated from the Russian by Nobumichi Mugibayashi and Alan Jeffrey. Pure and Applied Physics, Vol. 24 Academic Press, New York-London 1966.

[8] Biagini, F.; Hu, Y.Z.; Øksendal, B. and Sulem, A. A stochastic maximum principle for processes driven by fractional Brownian motion. Stochastic Process. Appl. 100 (2002), 233–253.

[9] El-Nouty, C. A Hanson-Russo-type law of the iterated logarithm for fractional Brownian motion. Statist. Probab. Lett. 17 (1993), no. 1, 27–34.

[10] Li, W. V. and Linde, W. Existence of small ball constants for fractional Brownian motions. C. R. Acad. Sci. Paris Sr. I Math. 326 (1998), no. 11, 1329–1334.

[11] Kuelbs, J. and Li, W. V. A functional LIL and some weighted occupation measure results for fractional Brownian motion. J. Theoret. Probab. 15 (2002), no. 4, 1007-1030.

[12] Bardi, M.; Crandall, M. G.; Evans, L. C.; Soner, H. M.; Souganidis, P. E. Viscosity solutions and applications. Lectures given at the 2nd C.I.M.E. Session held in Montecatini Terme, June 12–20, 1995. Edited by I. Capuzzo Dolcetta and P. L. Lions. Lecture Notes in Mathematics, 1660. Springer-Verlag, Berlin, 1997.

[13] Biagini F., Øksendal B. Hu Y.Z. and Sulem A. A stochastic maximum principle for processes driven by fractional Brownian motion. Stochastic processes and Applications, 100 (2002), 233-253.

[14] Biagini F., Øksendal B., Sulem A. and Wallner, N. An introduction to white noise theory and Malliavin calculus for fractional Brownian motion. Preprint, 2002.

[15] Borodin, A. N. and Salminen, P. Handbook of Brownian motion-facts and formulae. Second edition. Probability and its Applications. Birkhäuser Verlag, Basel, 2002.

[16] Bouleau, N. and Lépingle, D. Numerical methods for stochastic processes. Wiley Series in Probability and Mathematical Statistics: Applied Probability and Statistics. A Wiley-Interscience Publication. John Wiley & Sons, Inc., New York, 1994.

[17] Buckdahn, R. Anticipative Girsanov transformations and Skorohod stochastic differential equations. Mem. Amer. Math. Soc. 111 (1994), no. 533,

[18] Huang, S. T. and Cambanis, S. Stochastic and multiple Wiener integrals for Gaussian processes. Ann. Probab. 6 (1978), no. 4, 585-614.

[19] Carmona, P. and Coutin, L. Intégrale stochastique pour le mouvement brownien fractionnaire. C. R. Acad. Sci. Paris Sér. I Math. 330 (2000), no. 3, 231–236.

[20] Cheridito, P. Gaussian moving averaging, semimartingales and option pricing. Stoch. Proc. Appl. 109 (2004), 47–68.

[21] Cheridito P. and Nualart, D. Stochastic integral of divergence type with respect to fractional Brownian motion with Hurst parameter $H \in (0, \frac{1}{2})$. Preprint.

[22] Coutin, L. and Decreusefond, L. Stochastic Volterra equations with singular kernels. Stochastic analysis and mathematical physics, 39–50, Progr. Probab., 50, Birkhuser Boston, Boston, MA, 2001.

[23] Coutin, L., Nualart, D., Tudor, C.A. Tanaka formula for the fractional Brownian motion. Stochastic Processes Appl. 94 (2001), 301-315.

[24] Coutin, L. and Qian, Z.M. Stochastic analysis, rough path analysis and fractional Brownian motions. Probab. Theory Related Fields 122 (2002), no. 1, 108-140.

[25] Dai, W. and Heyde, C.C. Itô's formula with respect to fractional Brownian motion and its application. J. Appl. Math. Stoch. Anal. 9 (1996), 439-448.

[26] Dellacherie, C. and Meyer, P. A. Probabilités et potentiel. I, II, III, IV, V. Actualités Scientifiques et Industrielles Hermann, Paris, 1975, 1980, 1983, 1987, 1992.

[27] Dasgupta, A.; Kallianpur, G. Chaos decomposition of multiple fractional integrals and applications. Probab. Theory Related Fields 115 (1999), no. 4, 527–548.

[28] Dasgupta, A.; Kallianpur, G. Multiple fractional integrals. Probab. Theory Related Fields 115 (1999), 505-525.

[29] Delgado, R. and Sanz, M. The Hu-Meyer formula for nondeterministic kernels. Stochastics Stochastics Rep. 38 (1992), no. 3, 149–158.

[30] Decreusefond, L. A Skohorod-Stratonovitch integral for the fractional Brownian motion. Stochastic analysis and related topics, VII (Kusadasi, 1998), 177-198, Progr. Probab., 48, Birkhäuser Boston, Boston, MA, 2001.

[31] Decreusefond L. Calcul stochastique pour les processus de type Volterra. Thesis, 2001.

[32] Decreusefond, L. and Üstünel, A. S. Application du calcul des variations stochastiques au mouvement brownien fractionnaire. C. R. Acad. Sci. Paris Sér. I Math. 321 (1995), no. 12, 1605-1608.

[33] Decreusefond, L. and Üstünel, A.S. Stochastic analysis of the fractional Brownian motion. Potential Analysis 10 (1999), 177-214.

[34] Delgado, R. and Jolis, M. On a Ogawa-type integral with application to the fractional Brownian motion. Stochastic Anal. Appl. 18 (2000), no. 4, 617-634.

[35] Duncan, T.E.; Hu, Y.Z. and Pasik-Duncan, B. Stochastic calculus for fractional Brownian motion, I. Theory. SIAM Journal of Control Optimization, 38 (2000), 582-612.

[36] Doukhan P., Oppenheim G. and Taqqu M. S. Theory and applications of long-range dependence. Birkhäuser Boston, Inc., Boston, MA, 2003.

[37] Errami M. and Russo F. n-covariation, generalized Diricilet processes and calculus with respect to finite cubic variation processes. Preprint 2002.

[38] Elliott, R. J. van der Hoek, J. A general fractional white noise theory and applications to finance. Math. Finance 13 (2003), no. 2, 301–330.

[39] Fleming, W. H. and Soner, H. M. Controlled Markov processes and viscosity solutions. Applications of Mathematics, 25. Springer-Verlag, New York, 1993.

[40] Gaveau B. and Trauber, P. L'intégrale stochastique comme opérateur de divergence dans l'espace fonctionnel. J. Funct. Anal. 46 (1982), no. 2, 230-238.

[41] Glimm, J. Jaffe, A. Quantum physics. A functional integral point of view. Second edition. Springer-Verlag, New York, 1987.

[42] Gradinaru M. Nourdin I. Russo F. and Vallois P. n-order integrals and generalized Itô formula; the case of a fractional Brownian motion with any Hurst index. Preprint 2002.

[43] Gradinaru M. Russo F. and Vallois P. Generalized covariation, local time and Stratonovitch Itô formula for fractional Brownian motion with any Hurst index $H \geq 1/4$. Preprint 2001.

[44] Gripenberg G. and Norros I. On the Prediction of Fractional Brownian Motion. J. App. Prob. 33 (1996), 400-410.

[45] Guichardet, A. Symmetric Hilbert spaces and related topics. Infinitely divisible positive definite functions. Continuous products and tensor products. Gaussian and Poissonian stochastic processes. Lecture Notes in Mathematics, Vol. 261. Springer-Verlag, Berlin-New York, 1972.

[46] Holden, H.; Øksendal, B.; Ubøe, J. and Zhang, T.S. Stochastic partial differential equations. A modeling, white noise functional approach. Probability and its Applications. Birkhäuser Boston, Inc., Boston, MA, 1996.

[47] Hu, Y. Z. Prediction and translation of fractional Brownian motions. In Stochastics in Finite and Infinite Dimensions, T. Hida et al. editors, 153-171, Birkhäuser, 2000.

[48] Hu, Y. Z. Probability structure preserving and absolute continuity. Annales de l'Institut Henri Poincaré, 38 (2002), 557-580.

[49] Hu, Y.Z. Itô-Wiener chaos expansion with exact residual and correlation, variance inequalities. J. Theoret. Probab. 10 (1997), no. 4, 835–848.

[50] Hu, Y.Z. and Kallianpur, G. Exponential Integrability and Singular Infinite Dimensional Stochastic Differential Equations. Journal of Applied Mathematics and Optimization, 37 (1998), 295-353.

[51] Hu, Y.Z. and Kallianpur, G. and Xiong J. An approximation for Zakai equation. Appl. Math. Optim. 45 (2002), 23-44.

[52] Hu, Y.Z. and Meyer, P.A. Chaos de Wiener et Intégrales de Feynman. In Séminaire de Probabilités XXII, ed. by J. Azema, P.A. Meyer and M. Yor, Lecture Notes in Mathematics 1321, Springer-Verlag, 1988, 51-71.

[53] Hu, Y.Z. and Meyer, P. A. Sur les Intégrales Multiples de Stratonovich. In Séminaire de Probabilités XXVI, ed. by J. Azema, P.A. Meyer and M. Yor, Lecture Notes in Mathematics 1321, Springer-Verlag, 1988, 72-81.

[54] Hu, Y.Z. and Meyer, P.A. On the Approximation of Stratonovich Multiple Integrals. In Stochastic Processes, a festschrift in honor of G. Kallianpur, ed. by S. Cambanis, et al. 141-147, Springer, 1993.

[55] Hu, Y. and Nualart, D. Continuity of some anticipating integral processes. Statist. Probab. Lett. 37 (1998), no. 2, 203–211.

[56] Hu, Y. and Nualart, D. Some processes associated with fractional Bessel processes. Preprint, 2003.

[57] Hu, Y. and Nualart, D. Renormalized self-intersection local time for fractional Brownian motion. Preprint, 2003.

[58] Hu, Y.Z. and Øksendal, B. Fractional white noise calculus and applications to finance. Infin. Dimens. Anal. Quantum Probab. Relat. Top. 6 (2003), no. 1, 1–32.

[59] Hu, Y.Z. and Øksendal, B. Chaos expansion of local time of fractional Brownian motions. Stochastic Anal. Appl. 20 (2002), no. 4, 815–837.

[60] Hu, Y.Z. and Øksendal, B. Sulem, A. Optimal portfolio in a fractional Black & Scholes market. Mathematical physics and stochastic analysis (Lisbon, 1998), 267–279, World Sci. Publishing, River Edge, NJ, 2000.

[61] Hu, Y.; Øksendal, B. and Salopek, D. Weighted local time for fractional Brownian motion and applications to finance. Preprint, 2001.

[62] Hu, Y.Z. and Zhou, X.Y. Stochastic Control for Linear Systems Driven by Fractional Noises. Preprint 2002.

[63] Hult, H. Approximating some Volterra type stochastic integrals with applications to parameter estimation. Stochastic Process. Appl. 105 (2003), 1–32.

[64] Janson, S. Gaussian Hilbert spaces. Cambridge Tracts in Mathematics, 129. Cambridge University Press, Cambridge, 1997.

[65] Johnson, G. W. and Kallianpur, G. Some remarks on Y. Z. Hu and P.-A. Meyer's paper and infinite-dimensional calculus on finitely additive canonical Hilbert space. Theory Probab. Appl. 34 679–689 (1990)

[66] Johnson, G. W. and Kallianpur, G. Homogeneous chaos, p-forms, scaling and the Feynman integral. Trans. Amer. Math. Soc. 340 (1993), no. 2, 503–548.

[67] Jumarie, G. Stochastic differential equations with fractional Brownian motion input. Internat. J. Systems Sci. 24 (1993), no. 6, 1113-1131.

[68] I.S. Gradshteyn and I.M. Ryzhik, Table of Integrals, Series, and Products. San Diego : Academic Press, c2000.

[69] Kasahara, Y. and Matsumoto, Y. On Kallianpur-Robbins law for fractional Brownian motion. J. Math. Kyoto Univ. 36 (1996), no. 4, 815–824.

[70] Kôno, N. Kallianpur-Robbins law for fractional Brownian motion. Probability theory and mathematical statistics (Tokyo, 1995), 229–236, World Sci. Publishing, River Edge, NJ, 1996.

[71] Kusuoka, S. The nonlinear transformation of Gaussian measure on Banach space and absolute continuity. I. J. Fac. Sci. Univ. Tokyo Sect. IA Math. 29 (1982), no. 3, 567-597.

[72] Kusuoka, S. The nonlinear transformation of Gaussian measure on Banach space and its absolute continuity. II. J. Fac. Sci. Univ. Tokyo Sect. IA Math. 30 (1983), no. 1, 199-220.

[73] Levin, B. Ya. Lectures on entire functions. Translations of Mathematical Monographs, 150. American Mathematical Society, Providence, RI, 1996.

[74] Lin, S. J. Stochastic analysis of fractional Brownian motions. Stochastics and Stochastics Rep. 55 (1995), 121–140.

[75] Lobato, I. and Robinson, P. M. Averaged periodogram estimation of long memory. J. Econometrics 73 (1996), no. 1, 303–324.

[76] Lundgren, T. and Chiang, D. Solution of a class of singular integral equations. Quart. J. Appl. Math. 24, 303-313.

[77] Le Breton, A. Filtering and parameter estimation in a simple linear system driven by a fractional Brownian motion. Statist. Probab. Lett. 38 (1998), no. 3, 263-274.

[78] Lyons, T. J. Differential equations driven by rough signals. Rev. Mat. Iberoamericana 14 (1998), no. 2, 215–310.

[79] Lyons, T. and Qian, Z.M. System Control and Rough Paths. Clarendon Press. Oxford, 2002.

[80] Mandelbrot, B. B. and Taqqu, M. S. Robust R/S analysis of long-run serial correlation. Bull. Inst. Internat. Statist. 48 (1979), no. 2, 69–99.

[81] Mandelbrot, B.B. and Van Ness J.W. fractional Brownian motions, fractional noises and applications. SIAM Rev. 10 (1968) 422–437.

[82] Mémin, J.; Mishura, Y. and Valkeila, E. Inequalities for the moments of Wiener integrals with respect to a fractional Brownian motion. Statist. Probab. Lett. 51 (2001), 197-206.

[83] Meyer, P. A. Quantum probability for probabilists. Lect. Notes in Math. 1538, Springer, 1993.

[84] Meyer, P. A. Transformations de Riesz pour les lois gaussiennes. Seminar on probability, XVIII, 179-193, Lecture Notes in Math., 1059, Springer, Berlin, 1984.

[85] Métivier M. and Pellaumail, J. Stochastic Integration. Academic Press, New York-London-Toronto, 1980.

[86] Nualart, D. The Malliavin calculus and related topics. Probability and its Applications. Springer-Verlag, New York, 1995.

[87] Nualart, D.; Pardoux, E. Stochastic calculus with anticipating integrands. Probab. Theory Related Fields 78 (1988), 535-581.

[88] Nualart, D.; Răşcanu, A. Differential equations driven by fractional Brownian motion. Collect. Math. 53 (2002), no. 1, 55-81.

[89] Neveu, J. Processus aléatoires gaussiens. Les Presses de l'Université de Montréal, Montreal, Que. 1968.

[90] Norros, I.; Valkeila, E. and Virtamo, J. An elementary approach to a Girsanov formula and other analytic results on fractional Brownian motions. Bernoulli, 5 (1999), 571-587.

[91] Parthasarathy, K. R. and Schmidt, K. Positive definite kernels, continuous tensor products, and central limit theorems of probability theory. Lecture Notes in Mathematics, Vol. 272. Springer-Verlag, Berlin-New York, 1972.

[92] Pipiras, V. and Taqqu M. S. Integration questions related to fractional Brownian motion. Probab. Theory Related Fields 118 (2000), 251-291.

[93] Privault, N. Skorohod stochastic integration with respect to non-adapted processes on Wiener space. Stochastics Stochastics Rep. 65 (1998), no. 1-2, 13–39.

[94] Ramer, R. On nonlinear transformations of Gaussian measures. J. Functional Analysis 15 (1974), 166-187.

[95] Revuz, D. and Yor, M. Continuous martingales and Brownian motion. Third edition. Grundlehren der Mathematischen Wissenschaften, 293. Springer-Verlag, Berlin, 1999.

[96] Robinson, P. M. Log-periodogram regression of time series with long range dependence. Ann. Statist. 23 (1995), no. 3, 1048–1072.

[97] Russo, F. and Vallois, P. Stochastic calculus with respect to continuous finite quadratic variation processes. Stochastics Stochastics Rep. 70 (2000), no. 1-2, 1-40.

[98] Simon, B. The $P(\phi)_2$ Euclidean (quantum) field theory. Princeton Series in Physics. Princeton University Press, Princeton, N.J., 1974.

[99] Samko, S. G., Kilbas, A. A. and Marichev, O. I. Fractional Integrals and Derivatives, Theory and Applications. Gordon and Breach Science Publishers, 1993.

[100] Ševljakov, A. Ju. Itô's formula for an extended stochastic integral. Theory Probab. Math. Statist. No. 22 (1981), 163-174.

[101] Samorodnitsky, G. and Taqqu, M.S. Stable non-Gaussian random processes, stochastic models with infinite variance. Chapman & Hall, 1994.

[102] Solé, J. L. and Utzet, F. Stratonovich integral and trace. Stochastics Stochastics Rep. 29 (1990), no. 2, 203–220.

[103] Sugita, H. Hu-Meyer's multiple Stratonovich integral and essential continuity of multiple Wiener integral. Bull. Sci. Math. 113 (1989), no. 4, 463–474.

[104] Szabados, T. Strong approximation of fractional Brownian motion by moving averages of simple random walks. Stochastic Process. Appl. 92 (2001), 31-60.

[105] Szulga, J. and Molz, F. On simulating fractional Brownian motion. High dimensional probability, II (Seattle, WA, 1999), 377–387, Progr. Probab., 47, Birkhuser Boston, Boston, MA, 2000.

[106] Titchmarsh, E. C. The theory of functions. Oxford University Press, 1952.

[107] Talagrand, M. and Xiao, Y. Fractional Brownian motion and packing dimension. J. Theoret. Probab. 9 (1996), no. 3, 579–593.

[108] Tudor, C. Calcul stochastique anticipant et mouvement Brownien fractionnaire. Thèse de Doctorat de l'Université de la Rochelle, 2002.

[109] Üstünel, A. S. and Zakai, M. Transformation of measure on Wiener space. Springer Monographs in Mathematics. Springer-Verlag, Berlin, 2000.

[110] Xiao, Y. Hitting probabilities and polar sets for fractional Brownian motion. Stochastics Stochastics Rep. 66 (1999), 121-151.

[111] Xiao, Y. Packing dimension of the image of fractional Brownian motion. Statist. Probab. Lett. 33 (1997), no. 4, 379-387.

[112] Xiao, Y. Hölder conditions for the local times and the Hausdorff measure of the level sets of Gaussian random fields. Probab. Theory Related Fields 109 (1997), no. 1, 129–157.

[113] Yong, J. and Zhou, X. Y. Stochastic controls. Hamiltonian systems and HJB equations. Applications of Mathematics, 43. Springer-Verlag, New York, 1999.

[114] Yin, Z.-M. New methods for simulation of fractional Brownian motion. J. Comput. Phys. 127 (1996), no. 1, 66-72.

[115] Zähle, M. Integration with respect to fractal functions and stochastic calculus. I. Probab. Theory Related Fields 111 (1998), no. 3, 333-374.

[116] Zähle, M. Integration with respect to fractal functions and stochastic calculus. II. Math. Nachr. 225 (2001), 145-183.

Editorial Information

To be published in the *Memoirs*, a paper must be correct, new, nontrivial, and significant. Further, it must be well written and of interest to a substantial number of mathematicians. Piecemeal results, such as an inconclusive step toward an unproved major theorem or a minor variation on a known result, are in general not acceptable for publication. Papers appearing in *Memoirs* are generally longer than those appearing in *Transactions*, which shares the same editorial committee.

As of January 31, 2005, the backlog for this journal was approximately 5 volumes. This estimate is the result of dividing the number of manuscripts for this journal in the Providence office that have not yet gone to the printer on the above date by the average number of monographs per volume over the previous twelve months, reduced by the number of volumes published in four months (the time necessary for preparing a volume for the printer). (There are 6 volumes per year, each containing at least 4 numbers.)

A Consent to Publish and Copyright Agreement is required before a paper will be published in the *Memoirs*. After a paper is accepted for publication, the Providence office will send a Consent to Publish and Copyright Agreement to all authors of the paper. By submitting a paper to the *Memoirs*, authors certify that the results have not been submitted to nor are they under consideration for publication by another journal, conference proceedings, or similar publication.

Information for Authors

Memoirs are printed from camera copy fully prepared by the author. This means that the finished book will look exactly like the copy submitted.

The paper must contain a *descriptive title* and an *abstract* that summarizes the article in language suitable for workers in the general field (algebra, analysis, etc.). The *descriptive title* should be short, but informative; useless or vague phrases such as "some remarks about" or "concerning" should be avoided. The *abstract* should be at least one complete sentence, and at most 300 words. Included with the footnotes to the paper should be the 2000 *Mathematics Subject Classification* representing the primary and secondary subjects of the article. The classifications are accessible from www.ams.org/msc/. The list of classifications is also available in print starting with the 1999 annual index of *Mathematical Reviews*. The Mathematics Subject Classification footnote may be followed by a list of *key words and phrases* describing the subject matter of the article and taken from it. Journal abbreviations used in bibliographies are listed in the latest *Mathematical Reviews* annual index. The series abbreviations are also accessible from www.ams.org/publications/. To help in preparing and verifying references, the AMS offers MR Lookup, a Reference Tool for Linking, at www.ams.org/mrlookup/. When the manuscript is submitted, authors should supply the editor with electronic addresses if available. These will be printed after the postal address at the end of the article.

Electronically prepared manuscripts. The AMS encourages electronically prepared manuscripts, with a strong preference for $\mathcal{A}_{\mathcal{M}}\mathcal{S}$-LaTeX. To this end, the Society has prepared $\mathcal{A}_{\mathcal{M}}\mathcal{S}$-LaTeX author packages for each AMS publication. Author packages include instructions for preparing electronic manuscripts, the *AMS Author Handbook*, samples, and a style file that generates the particular design specifications of that publication series. Though $\mathcal{A}_{\mathcal{M}}\mathcal{S}$-LaTeX is the highly preferred format of TeX, author packages are also available in $\mathcal{A}_{\mathcal{M}}\mathcal{S}$-TeX.

Authors may retrieve an author package from e-MATH starting from `www.ams.org/tex/` or via FTP to `ftp.ams.org` (login as `anonymous`, enter username as password, and type `cd pub/author-info`). The *AMS Author Handbook* and the *Instruction Manual* are available in PDF format following the author packages link from `www.ams.org/tex/`. The author package can be obtained free of charge by sending email to `pub@ams.org` (Internet) or from the Publication Division, American Mathematical Society, 201 Charles St., Providence, RI 02904, USA. When requesting an author package, please specify $\mathcal{AMS}$-LaTeX or $\mathcal{AMS}$-TeX, Macintosh or IBM (3.5) format, and the publication in which your paper will appear. Please be sure to include your complete mailing address.

Sending electronic files. After acceptance, the source file(s) should be sent to the Providence office (this includes any TeX source file, any graphics files, and the DVI or PostScript file).

Before sending the source file, be sure you have proofread your paper carefully. The files you send must be the EXACT files used to generate the proof copy that was accepted for publication. For all publications, authors are required to send a printed copy of their paper, which exactly matches the copy approved for publication, along with any graphics that will appear in the paper.

TeX files may be submitted by email, FTP, or on diskette. The DVI file(s) and PostScript files should be submitted only by FTP or on diskette unless they are encoded properly to submit through email. (DVI files are binary and PostScript files tend to be very large.)

Electronically prepared manuscripts can be sent via email to `pub-submit@ams.org` (Internet). The subject line of the message should include the publication code to identify it as a Memoir. TeX source files, DVI files, and PostScript files can be transferred over the Internet by FTP to the Internet node `e-math.ams.org` (130.44.1.100).

Electronic graphics. Comprehensive instructions on preparing graphics are available at `www.ams.org/jourhtml/graphics.html`. A few of the major requirements are given here.

Submit files for graphics as EPS (Encapsulated PostScript) files. This includes graphics originated via a graphics application as well as scanned photographs or other computer-generated images. If this is not possible, TIFF files are acceptable as long as they can be opened in Adobe Photoshop or Illustrator. No matter what method was used to produce the graphic, it is necessary to provide a paper copy to the AMS.

Authors using graphics packages for the creation of electronic art should also avoid the use of any lines thinner than 0.5 points in width. Many graphics packages allow the user to specify a "hairline" for a very thin line. Hairlines often look acceptable when proofed on a typical laser printer. However, when produced on a high-resolution laser imagesetter, hairlines become nearly invisible and will be lost entirely in the final printing process.

Screens should be set to values between 15% and 85%. Screens which fall outside of this range are too light or too dark to print correctly. Variations of screens within a graphic should be no less than 10%.

Inquiries. Any inquiries concerning a paper that has been accepted for publication should be sent directly to the Electronic Prepress Department, American Mathematical Society, 201 Charles St., Providence, RI 02904, USA.

Titles in This Series

TITLES IN THIS SERIES

For a complete list of titles in this series, visit the
AMS Bookstore at **www.ams.org/bookstore/**.

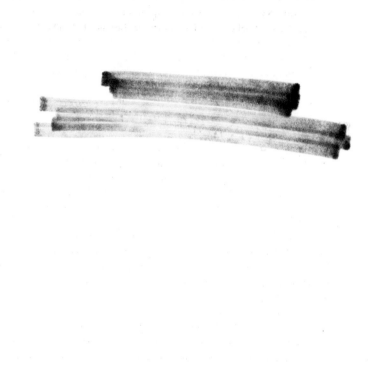

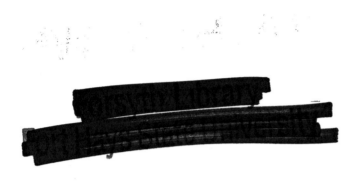